Evaluating Human Genetic Diversity

Committee on Human Genome Diversity

Commission on Life Sciences

NATIONAL RESEARCH COUNCIL

WITHDRAWN

NATIONAL ACADEMY PRESS
Washington, DC 1997

NATIONAL ACADEMY PRESS • 2101 Constitution Ave., N.W. • Washington, D.C. 20418

NOTICE: The project that is the subject of this report was approved by the Governing Board of the National Research Council, whose members are drawn from the councils of the National Academy of Sciences, the National Academy of Engineering, and the Institute of Medicine. The members of the committee responsible for the report were chosen for their special competences and with regard for appropriate balance. In preparing its report, the committee invited people with different perspectives to present their views. Such invitation does not imply endorsement of those views.

This report has been reviewed by a group other than the authors according to procedures approved by a Report Review Committee consisting of members of the National Academy of Sciences, the National Academy of Engineering, and the Institute of Medicine.

This study by the National Research Council's Commission on Life Sciences was sponsored by the National Institutes of Health and the National Science Foundation under contract no. N01-OD-4-2139. Points of view in this document are those of the authors and do not necessarily represent the official position of the sponsoring agencies.

Library of Congress Catalog Card Number 97-81059
International Standard Book Number 0-309-05931-3

Additional copies of this report are available from:
National Academy Press
2101 Constitution Ave., NW
Box 285
Washington, DC 20055
800-624-6242
202-334-3313 (in the Washington Metropolitan Area)
http://www.nap.edu

Copyright 1997 by the National Academy of Sciences. All rights reserved.

Printed in the United States of America

COMMITTEE ON HUMAN GENOME DIVERSITY

WILLIAM J. SCHULL (*Chair*) University of Texas Health Center, Houston, TX
GEORGE J. ANNAS, Boston University Schools of Medicine and Public Health, Boston, MA
NORMAN ARNHEIM, University of Southern California, Los Angeles, CA
JOHN BLANGERO, Southwest Foundation for Biomedical Research, San Antonio, TX
ARAVINDA CHAKRAVARTI, Case Western Reserve University, Cleveland, OH
VIRGINIA R. DOMINGUEZ, University of Iowa, Iowa City, IA
GEORGIA DUNSTON, Howard University, Washington, DC
WARD H. GOODENOUGH, University of Pennsylvania, Philadelphia, PA
RICHARD R. HUDSON, University of California, Irvine, CA
ERIC JUENGST, Case Western Reserve University, Cleveland, OH
MICHAEL M. KABACK, University of California, San Diego, CA
DANIEL R. MASYS, University of California, San Diego, CA
KATHRYN MOSELEY, Henry Ford Hospital, Detroit, MI
ROBERT SOKAL, State University of New York, Stony Brook, NY
ALAN R. TEMPLETON, Washington University, St. Louis, MO
LAP-CHEE TSUI, The Hospital for Sick Children, Toronto, ON, Canada
GEORGE C. WILLIAMS, State University of New York, Stony Brook, NY

NRC Staff

TANIA WILLIAMS, Study Director
ERIC A. FISCHER, Study Director (through December 1996)
NORMAN GROSSBLATT, Editor
ERIKA SCHUGART, Research Assistant
KATHLEEN BEIL, Project Assistant
PAULETTE A. ADAMS, Senior Project Assistant (through August 1996)

BOARD ON BIOLOGY

MICHAEL T. CLEGG (*Chair*) University of California, Riverside, CA
JOHN C. AVISE, University of Georgia, Athens, GA
DAVID EISENBERG, University of California, Los Angeles, CA
GERALD D. FISCHBACH, Harvard Medical School, Cambridge, MA
DAVID J. GALAS, Darwin Molecular Corporation, Bothell, WA
DAVID GOEDDEL, Tularik, Incorporated, South San Francisco, CA
ARTURO GOMEZ-POMPA, University of California, Riverside, CA
COREY S. GOODMAN, University of California, Berkeley, CA
BRUCE R. LEVIN, Emory University, Atlanta, GA
OLGA F. LINARES, Smithsonian Tropical Research Institute, Panama
ELLIOTT M. MEYEROWITZ, California Institute of Technology, Pasadena, CA
ROBERT T. PAINE, University of Washington, Seattle, WA
COREY S. GOODMAN, University of California, Berkeley, CA
RONALD R. SEDEROFF, North Carolina State University, Raleigh, NC
DANIEL SIMBERLOFF, Florida State University, Tallahassee, FL
ROBERT R. SOKAL, State University of New York, Stony Brook, NY
SHIRLEY TILGHMAN, Princeton University, Princeton, NJ
RAYMOND L. WHITE, University of Utah, Salt Lake City, UT

Staff

PAUL GILMAN, Acting Director
ERIC A. FISCHER, Director (through December 1996)
TANIA WILLIAMS, Program Officer
KATHLEEN BEIL, Administrative Assistant
ERIKA SHUGART, Research Assistant

COMMISSION ON LIFE SCIENCES

THOMAS D. POLLARD (*Chair*) The Salk Institute for Biological Studies, La Jolla, CA
FREDERICK R. ANDERSON, Cadwalader, Wickersham & Taft, Washington, DC
JOHN C. BAILAR, III, University of Chicago, IL
PAUL BERG, Stanford University, Stanford, CA
JOANNA BURGER, Rutgers University, Piscataway, NJ
SHARON L. DUNWOODY, University of Wisconsin, Madison, WI
JOHN L. EMMERSON, Indianapolis, IA
NEAL L. FIRST, University of Wisconsin, Madison, WI
URSULA W. GOODENOUGH, Washington University, St. Louis, MO
HENRY W. HEIKKINEN, University of Northern Colorado, Greeley, CO
HANS J. KENDE, Michigan State University, East Lansing, MI
CYNTHIA J. KENYON, University of California, San Francisco, CA
DAVID M. LIVINGSTON, Dana-Farber Cancer Institute, Washington, DC
DONALD R. MATTISON, University of Pittsburgh, Pittsburgh, PA
JOSEPH E. MURRAY, Wellesley Hills, MA
EDWARD E. PENHOET, Chiron Corporation, Emeryville, CA
MALCOLM C. PIKE, Norris/USC Comprehensive Cancer Center, Los Angeles, CA
JONATHAN M. SAMET, The Johns Hopkins University, Baltimore, MD
CHARLES F. STEVENS, The Salk Institute for Biological Studies, La Jolla, CA
JOHN L. VANDEBERG, Southwest Foundation for Biomedical Research, San Antonio, TX

Staff

PAUL GILMAN, Executive Director
SOLVEIG PADILLA, Administrative Assistant

The National Academy of Sciences is a private, nonprofit, self-perpetuating society of distinguished scholars engaged in scientific and engineering research, dedicated to the furtherance of science and technology and to their use for the general welfare. Upon the authority of the charter granted to it by the Congress in 1863, the Academy has a mandate that requires it to advise the federal government on scientific and technical matters. Dr. Bruce Alberts is president of the National Academy of Sciences.

The National Academy of Engineering was established in 1964, under the charter of the National Academy of Sciences, as a parallel organization of outstanding engineers. It is autonomous in its administration and in the selection of its members, sharing with the National Academy of Sciences the responsibility for advising the federal government. The National Academy of Engineering also sponsors engineering programs aimed at meeting national needs, encourages education and research, and recognizes the superior achievements of engineers. Dr. William Wulf is president of the National Academy of Engineering.

The Institute of Medicine was established in 1970 by the National Academy of Sciences to secure the services of eminent members of appropriate professions in the examination of policy matters pertaining to the health of the public. The Institute acts under the responsibility given to the National Academy of Sciences by its congressional charter to be an adviser to the federal government and, upon its own initiative, to identify issues of medical care, research, and education. Dr. Kenneth I. Shine is president of the Institute of Medicine.

The National Research Council was organized by the National Academy of Sciences in 1916 to associate the broad community of science and technology with the Academy's purposes of furthering knowledge and advising the federal government. Functioning in accordance with general policies determined by the Academy, the Council has become the principal operating agency of both the National Academy of Sciences and the National Academy of Engineering in providing services to the government, the public, and the scientific and engineering communities. The Council is administered jointly by both Academies and the Institute of Medicine. Dr. Bruce Alberts and Dr. William A. Wulf are chairman and vice chairman, respectively, of the National Research Council.

Preface

Over the last three-fourths of this century or so, an enormous body of data has accumulated on the extent of genetic variation among human beings, but most of this information has arisen opportunistically, having been driven by individual investigator initiatives and collected under widely varied conditions. Moreover, the information and samples that have been collected are dispersed in laboratories around the world, and access to them is often difficult. Therefore, it has proved difficult to compare results from different studies, and this difficulty has narrowed the value of the information and samples for the study of many problems of current evolutionary and biologic interest. To remedy those shortcomings, support has been growing in the international scientific community for a worldwide, geographically comprehensive survey of variation in the human genome.

The Committee on Human Genome Diversity, in the Board on Biology of the National Research Council's Commission on Life Sciences, came into being as a result of a request from the National Science Foundation and the National Institutes of Health for the Research Council to assess the scientific value, technical aspects, and organizational requirements of a systematic worldwide survey of human genetic variability and the ethical, legal, and social issues that would be raised by it before the commitment of substantial funds to any survey. The committee was organized in early 1996 and was structured to include members with expertise in all the major fields relevant to the project: population, human, and molecular genetics; evolutionary biology; anthropology (cultural and biologic); biostatistics; informatics; ethics; and law.

In its fact-finding, it became apparent to the committee that the precise nature of the proposed survey was more elusive than the committee had envisioned; different participants in the formulation of its consensus document had quite different perceptions of the intent of the project and even of its organizational structure. The committee reviewed the consensus document for the proposed Human Genome Diversity Project (HGDP) and was briefed by persons involved in formulating it. The committee found that there was no sharply defined proposal that it could evaluate; as a result, it chose to look at the issues posed by such a global survey of human genetic variation more broadly.

The committee held 4 meetings; at 3, members of the public and persons acting on the public's behalf were invited to discuss the issues with the committee, whereas the fourth was devoted entirely to the writing of this report. To elicit as wide a spectrum of opinions on the merits of the proposed survey as practical, the committee circulated a questionnaire encouraging those who could not attend the public sessions to submit their opinions in writing. Their comments were tabulated and taken into account in the committee's deliberations and this report.

The committee gratefully acknowledges the support of staff of the National Research Council. Eric Fischer and Tania Williams helped to refine the report and contributed to the preparation and administrative work of the project; Norman Grossblatt edited the manuscript; and invaluable support was provided by Kathleen Beil, Erika Shugart, and Paulette Adams.

Contents

EXECUTIVE SUMMARY   1
   Sampling Issues, 2
   Sample Collection and Data Management, 5
   Human Rights Considerations, 6
   Organization and Management, 8

1 INTRODUCTION AND BACKGROUND   10
   The Committee, Its Charge, and Its Activities, 11
   The Proposed Human Genome Diversity Project, 12
   This Report, 14

2 SCIENTIFIC AND MEDICAL VALUE OF
   RESEARCH ON HUMAN GENETIC VARIATION   16
   Human History and Evolution, 17
   Basic Mechanisms of Genome Evolution, 19
   Biomedical Applications, 20
   Conclusion, 22

3 SAMPLING ISSUES   23
   Basic Sampling Strategies, 24
   Strategies That Are Not Population Based, 25
   Strategies That Are Population Based, 25
   Which Sampling Strategy Should Be Used?, 30
   How to Select Which Human Populations to Sample, 32

Considerations in Choosing Subject Populations, 32
Number of Populations to Sample, and Number of People
to Sample in a Population, 33
Summary and Conclusions, 35

4 SAMPLE COLLECTION AND DATA MANAGEMENT   36
Sources of DNA to Be Sampled, 36
Characterization of Genetic Variation, 39
Classes of DNA Markers, 40
DNA Polymorphisms Based on Single Nucleotide Substitutions, 42
Should There Be a Core Set of DNA Markers That Will Be Scored
for All Samples in the Repository?, 43
Research-Materials Management, 44
Data Management, 47
Functional Requirements for Human Genetic Variation Data, 49
Data Acquisition Methods, 50
Quality Control and Annotation, 50
Communication Via Public Networks, 51
Archival Storage, 51
Distribution and Access, 52
Security Issues Related to Human Genetic Variation Data, 52
Summary and Conclusions, 53

5 HUMAN RIGHTS AND HUMAN GENETIC-
VARIATION RESEARCH   55
Context of Concerns About Studying Human Genetic Variation, 56
Ethical Considerations in the Design of Human
Genetic-Variation Research, 58
Control, 65
Commercialization and Reciprocity Agreements, 66
Conclusion, 68

6 ORGANIZATIONAL AND OTHER ISSUES   69

REFERENCES   74

APPENDIX A: Committee on Human Genome Diversity:
Biographical Information   81

APPENDIX B: Acknowledgments   89

Executive Summary

At the request of the National Science Foundation and the National Institutes of Health, the Committee on Human Genome Diversity was organized in 1996 in the Board of Biology of the National Research Council's Commission on Life Sciences. The committee's charge, as defined in the agreement between the Research Council and the sponsors, was to evaluate the consensus proposal to establish a Human Genome Diversity Project (HGDP).

As the committee's fact-finding progressed, it became apparent that the precise nature of the proposed HGDP was elusive; different participants in the formulation of the "consensus" document had quite different perceptions of the intent of the project, and even of its organizational structure. Accordingly, because there was no sharply defined proposal that the committee could evaluate, it chose to examine the scientific merits and value of research on human genetic variation and the organizational, policy, and ethical issues that such research poses in a more-general context.

The committee, which comprised representatives of all the relevant disciplines, met on 4 occasions to respond to its charge. At these meetings, spokespersons for the scientific community and the public were invited to share their views with the committee. To provide an opportunity for those unable to present their opinions in person, the committee circulated a questionnaire to numerous individuals and organizations. It received hundreds of responses to the questionnaire, for which the committee is most appreciative. The report that follows considered those oral and written presentations.

Briefly, the committee is persuaded that a global assessment of the extent of human genetic variability has substantial scientific merit and warrants support, largely because of the insight that the data collected could provide into the origin and evolution of the human species. A comprehensive survey of human genetic variability both between and within populations could map such variability and place it in social and environmental context. Careful variability sampling in conjunction with the Human Genome Project could contribute fundamentally to a new era of modern molecular medicine and transform scientific understanding of human evolution and the course of human prehistory.

However, the committee foresees numerous ethical, legal, and human-rights challenges in the prosecution of a global effort and offers possible guidelines to the resolution of some, albeit not all, of the challenges that the committee identifies. The committee recognizes that these ethical, legal, and human-rights concerns will not be the only problems. A global survey will also pose numerous technical and organizational difficulties, of which some can be foreseen now and others cannot. Those difficulties, although complex, appear more tractable than do some of the ethical, legal, and moral ones, primarily because solutions to the latter must embrace a wide array of different value systems and cultures, each with its own sense of the rights and obligations of individuals and of the group to which they belong. The committee offers suggestions and recommendations on sampling, specimen collection and storage, and data management that in its view would mitigate some of the technical and organizational difficulties inherent in a multinational, multicultural study.

SAMPLING ISSUES

The committee considered 5 possible sampling strategies (see table 1) to determine which one would be most appropriate for a coordinated global effort.

Strategy I is the simplest sampling scheme because its sole requirement is that the sample be representative of the human species. To achieve this end the sample should not be derived from one restricted group of human beings. This scheme yields a sample that cannot be linked to specific individuals, geographic areas or populations. Each sample is identified simply as being from a human being, and no other information is obtained. This is the least-expensive type of sample to acquire, and its collection minimizes many ethical issues at both the personal and population levels (see chapter 5).

Strategy II differs from Strategy I in that it records the geographic location of each sampling point but the sample cannot be linked to particular persons or populations. The geographic points to be sampled could be chosen either by using a grid method or by sampling geographic areas in proportion to the density of populations in them. All the hypotheses related to genome evolution and patterns of variation that were testable with strategy I are also testable with strategy II, but in addition it is possible to test hypotheses related to the patterns of spatial variation and some hypotheses about the geographic subdivision of humans and patterns of gene flow or migration.

TABLE 1 Sampling Strategies

	Non-Population-Based Sampling		Population-Based Sampling		
	I Totally Anonymous	II Geographic Location	III II + Group Identification Data	IV III + Individual Phenotypic Data	V IV + Pedigrees
Testable Hypotheses	Genome evolution Patterns of variation in the genome Overall genetic variation in humans	Same as I plus: Description and determination of spatial variation (such as, variation of loci in space, migration)	Same as II plus: Patterns of migration, gene flow, and population subdivision Hypotheses from anthropology, archaeology, history, and linguistics that should affect patterns of interpopulation variation Preliminary studies on medically relevant loci Population-level medical associations	Same as III plus: Identify specific loci for possible biomedical applications Genotype interactions Within-group variation on medical and phenotypic data Associations between genes and phenotype at individual level	Same as IV plus: Detailed studies on disease-associated genes
Relative Costs	$	$$	$$$	$$$$	$$$$$

Strategy III is the first of the 3 population-based sampling designs given in table 1. It records not only the geographic location of a sample, but also information provided about self-reported ethnicity, primary language, sex, age, and parental birthplaces. However, no personal identifiers would be obtained and thus neither the information nor the samples could be linked to specific persons. All the hypotheses that are testable with strategies I and II can be tested with strategy III, but this strategy broadens the universe of testable hypotheses to those related to population-level relationships and differences measured primarily with data on the frequencies of alleles (alternative forms of genes at the same locus) or haplotypes (particular states of a region of DNA if the DNA region is a coding region, haplotypes correspond to alleles), and associated and derived statistics.

Strategy IV, the second of the population-based strategies given in table 1, includes biomedically relevant information on individually identifiable phenotypes, particularly disease phenotypes. All the hypotheses mentioned in connection with strategy III could be tested with this scheme, but in addition one could look for genotype-disease associations instead of the much-weaker population-disease associations possible with strategy III. However, even such an enhanced data set would still be limited to disease-association studies and could not address disease causation directly. Hidden or unknown heterogeneity in the populations sampled could easily lead to false conclusions, and additional sampling (often the gathering of pedigree data) would be needed to confirm the results obtained with this strategy.

Those limitations can be avoided by going to a third level of population sampling, strategy V, the sampling of families or pedigreed persons in a population instead of persons of unknown relationship. When pedigree data are gathered with population and phenotypic data, more-definitive phenotypic studies are possible and they have enhanced power to detect markers close enough to disease loci to produce a within-family association. Moreover, when many closely linked marker loci exhibit heterozygosity, family data often allow the construction of haplotypes with more certainty. Therefore, this form of sampling would greatly increase the biomedical utility of a human genome sample collection.

Of the various sampling strategies discussed and summarized in table 1, population-based sampling strategy III, in which only basic group-identification data are gathered, is recommended over the other strategies since the data and specimens cannot be linked to specific individuals. Strategy I does not provide a rationale for global sampling, and strategy II has many of the same ethical complications as strategy III but with a substantial restriction in breadth of testable hypotheses. Strategies IV and V could greatly increase the cost, complicate sampling logistics, raise serious ethical and security concerns, and benefit only a few investigators (although the investigations that would be so benefited have the most-direct biomedical relevance). Strategy III offers the best balance of breadth of testable hypotheses, expense, and ethical complications.

A coordinated global sampling effort to develop a common resource for research on human genome variation should use a population-based sampling design in which the geographic location of the sample and self-reported ethnicity, primary language, sex, age, and parental birthplaces are recorded. The committee notes that the inclusion of parental birthplaces along with the other information identified above could, in some instances, inadvertently identify specific individuals.

For any given population, samples of a few hundred to several hundred persons, or even more, should be obtained whenever possible. In larger populations where the investigator(s) deems stratified sampling to be necessary, larger overall samples would be desirable.

SAMPLE COLLECTION AND DATA MANAGEMENT

The committee believes that at a stage when genotyping technology is evolving rapidly, it would be scientifically inappropriate and premature to designate a common core set of markers that is to be genotyped in all samples. Given advances in technology, a natural outcome will be that individual investigators will perform large-scale surveys of a large number of markers to generate balanced data sets. In spite of differences among individual investigators in sampling designs due to the different hypotheses being tested, many will use common technologies that can provide uniformity in the types and numbers of markers analyzed. To encourage and to facilitate as widespread testing as possible, blood samples collected from human populations should be converted primarily into purified DNA.

With currently available laboratory and information technologies, the material-management and data-management aspects of a coordinated human genome variation research effort do not appear to constitute a serious barrier to implementation of the project. There are multiple feasible models for specimen and data management and numerous instances of international cooperation in the creation of shared repositories of biologic tissue and data. The specimens and data to be captured, analyzed, and disseminated by the project have unique aspects, which will require attention and resources, but none of them is intractable. The most-important decision about project design will be whether it will acquire specimens and data that can be linked to specific individuals and thereby need to meet a "clinical" standard for specimen and data security and access control.

The committee was not charged with providing detailed guidance on data management; therefore our remarks are directed toward general issues. However, there are numerous questions that need to be addressed in a more specific manner than the committee believed itself charged to do. These details must be resolved before embarking on a major data collecting enterprise: The committee recommends that a special panel be convened to do so. Among these questions are the following:

- Is the data system to be a static (add-only) archive, an originator-modifiable archive, or a cooperative work system? Will the system store results of analyses of the primary data, or third-party annotations, or comments on the basic data?
- How will release to public access be handled? Will there be multiple levels of access?
- How will data-sample links be assured?
- If the database is to be accessible to members of the participating populations, how will the multi-language interfaces that such would require be developed?
- How will maintenance be done and funded over the long term?
- How will the databases be structured to meet the conflicting ends of archiving and use of data?

Other important considerations are the following:

- Establishment of a resource-allocation mechanism to monitor and adjudicate requests for both renewable and nonrenewable research materials,
- A review mechanism for determining the scientific and ethical merit of requests for specimens (analogous to an institutional review board),
- A mechanism to detect and respond to unauthorized reuse of specimens for research not agreed to by subject populations,
- If individually identifiable specimens are collected, a procedure the committee does not advocate, then a mechanism must be established for recontact with and reconsent of participating groups and persons if currently unforeseen uses of specimens arise that are beyond the scope of the original informed consent.
- Enforcement of ethical protocols, especially the right of groups and individual persons to withdraw their samples from the collection if the samples are personally identifiable.

HUMAN RIGHTS CONSIDERATIONS

Collecting biologic samples from specific individuals and families to extrapolate information about the social groups to which they belong is not a new scientific practice. The confluence of several sets of ethical considerations gives that practice greater risks that human genetic variation researchers must recognize. Continued use of outmoded social categories to structure biomedical research, emerging possibilities for commercializing biomedical knowledge, and heightened awareness of the stigmatizing potential of genetic information all increase public concern about human genetic variation research. To the extent that such research must continue to rely on socially defined human groups, the process of managing any coordinated effort to survey human diversity will be increasingly complex. For each socially identified set of samples, protocols for group involvement and concurrence (including in the design of the research protocol) will have to be negotiated and balanced against the researchers' fundamental ethical obligations to protect the freedom, privacy, and welfare of the individuals involved, including the right not to participate in a study.

It is crucial to have a complete research protocol for review before the actual consent form and process for obtaining consent can be designed and evaluated. For any specific goal-oriented protocol, it should be possible to anticipate the risks and benefits to the subjects and pursue informed consent accordingly. For projects that are not able to specify goals in sufficient detail to quantify risks and benefits reasonably, the worst-case scenario should be assumed: the benefits will be at the lowest anticipated level, and the risks at the highest. That means that the burden of proof for any DNA-sampling project that does not have a well-defined hypothesis will be high. It also underlines the most basic starting point for all ethical analyses of genetic-variation research, regardless of which model is pursued: defining a hypothesis and determining the benefit of knowing whether it is true.

Accurate identification of population units for sampling purposes requires extensive knowledge of the social, political, and linguistic composition of the region to be sampled. Published ethnographic studies can provide some of this knowledge, as can anthropologists who work with the peoples. If this information is not available, researchers should study the local situation in consultation with local leaders, experts, and other researchers before designing the sampling strategy.

In locations where women's rights to self-determination are not recognized (and thus their informed consent not possible), "women should not normally be involved in the research" (commentary on guideline 11 of the International Ethical Guidelines for Biomedical Research Involving Human Subjects), because it is likely that they will not have the freedom and power to choose whether to participate. While it is obviously wrong to exclude women from participation in a study that could lead to results from which they could benefit, it is equally important to insist on informed consent that is freely given.

We think that it is too extreme a position always to require both group and individual consent to DNA collection for genetic-variation research. Nonetheless, researchers will have to make sure that their participants understand both the objections of their community and the rationale for them as part of the informed-consent process and, when doing research that is opposed by a specific community, will also have to take into account the possible impact of doing such research on the likelihood that other communities will cooperate with other genetic-variation researchers in the future.

Should the population itself be able to withdraw from the project? The answer might be that "community withdrawal" is not possible; if that is the case, it should be spelled out in both the protocol and the individual consent processes, as well as in the discussion of the protocol with community representatives. In general, consent and withdrawal are rights of individual research subjects and should not depend on the approval or disapproval of government authorities, however defined.

Studies that collect DNA specimens that can be linked to specific, identifiable persons must institute measures that will prevent unauthorized access to this information, so as to prevent individual research participants from stigmatization and discrimination, and must include mechanisms for follow-up about the results of the studies conducted on collected samples. It is not ethically or legally acceptable to ask research participants to "consent" to future but yet-unknown uses of their identifiable DNA samples. Consent in such a case is a waiver of rights, and such waivers are explicitly prohibited by federal research regulations.

Arrangements regarding financial interests in the products or outcomes of the research should be negotiated as part of the original project review and informed-consent process. In addition, a monitoring and enforcement mechanism, with representation of the affected groups, should be in place. One of the major lessons from the Rio de Janeiro Biodiversity Summit is the importance of economic and political considerations in negotiating research participation with identified human groups. That should not be surprising, inasmuch as social groups are usually created and sustained as a means of pursuing their members' economic and political interests. However, this adds a dimension to informed-consent negotiations that is foreign to most social and biomedical scientists: negotiating over what the participating group receives in return for participation.

ORGANIZATION AND MANAGEMENT

We recognize that neither the National Science Foundation nor the National Institutes of Health are prepared or even able to fund a global survey such as that contemplated and that they seek advice on the role they should play. Accordingly, the committee offers the following guidance: These agencies should focus their financial support, at least initially, on projects originating in the United States and expand their support to the international scene only after the US activities are successfully launched. The establishment of an international effort will require defining the roles of interested investigators, on the one hand, and national and international agencies, on the other. Without defining such roles, any global survey would be correctly criticized for substituting a self-appointed set of administrators without official standing in any country for the recognized national and international agencies of governance, and is unlikely to succeed. The funding agencies, specifically the National Science Foundation and the National Institutes of Health, should initiate such discussions through their international offices. These discussions will take time to bring to fruition, and until a consensus is achieved the US effort would be generating information of substantial moment relevant to the feasibility and urgency of an international study, and identifying administrative barriers that would have to be surmounted.

The committee found its deliberations on the value, design and implementation of tissue repositories, whether centrally or regionally located, constantly thwarted by the absence of information on what repositories are actually available now and the specimens that might be accessible to other investigators. Such information would be of substantial use to many in the scientific community. The committee recommends that NIH or NSF identify all such repositories as well as the availability of the specimens to the scientific community in the United States as well as elsewhere.

Finally, the recommendations of this committee with regard to sampling strategy, sample size, and the collection of specimens and data should be taken into account when considering the scientific merit of an individual request for support.

1

Introduction and Background

Genetic variation is the material basis of evolution, and its origin and maintenance have long been the focus of population genetics and evolutionary biology. Charles Darwin (1859) recognized the importance of variation in the origin of species long before modern genetics began to emerge, and numerous other early investigators noted that human groups differ in skin color, hair form, body size, and other physical characteristics. However, it was the Hirschfelds, in a series of seminal publications shortly after World War I (see, for example, Hirschfeld and Hirschfeld 1919), who first established that human populations differ substantially in the frequency of specific, simply inherited genetic traits, namely, the ABO blood types. Later work has shown that such variation can be measured at cellular and molecular levels, as in the red-cell and white-cell antigens, serum proteins, leukocyte enzymes, mitochondrial DNA, and genomic DNA itself (for compilations of the observed variability, see Cavalli-Sforza and others 1994; Mourant and others 1976; Roychoudury and Nei 1988). The variation has been used to explore the history of human groups, to study the evolutionary relationships among existing nonhuman primates and humans, to estimate the importance of selection and genetic drift in the shaping of the genetic structure of populations, and to determine the time of origin and the spread of specific mutations.

An enormous body of data has accumulated on the extent of genetic variation among human beings, but most of the information has arisen opportunistically, having been driven by individual investigator initiatives and collected under widely differing conditions. Moreover, the information and samples that have been collected are dispersed in laboratories around the world, and access to them

INTRODUCTION AND BACKGROUND

is often limited. Therefore it has proved difficult to compare results from different studies and this difficulty has limited the value of the information and samples for the study of many problems of current evolutionary and biomedical interest. To remedy those shortcomings, support has been growing in the international scientific community for a worldwide, geographically comprehensive survey of variation in the human genome. The most well-developed and widely recognized proposal for conducting such a survey is known as the Human Genome Diversity Project (HGDP).

THE COMMITTEE, ITS CHARGE, AND ITS ACTIVITIES

The Committee on Human Genome Diversity of the Board on Biology of the National Research Council's Commission on Life Sciences came into being as a result of a request from the National Science Foundation (NSF) and the National Institutes of Health (NIH) for the Research Council to assess the scientific value, technical aspects, and organizational requirements of a systematic worldwide survey of human genetic variability—and the ethical, legal and social issues that would be raised by it—before the commitment of any substantial funds to any such survey. The committee was organized in early 1996 and was structured to include members with expertise in all the major fields relevant to the project: population, human, and molecular genetics; evolutionary biology; anthropology (cultural and biologic); biostatistics; informatics; ethics; and law.

The committee's charge, as defined in the agreement between the Research Council and the sponsors, was to evaluate the proposal to establish the HGDP. The evaluation was to include examination of the scope and objectives of the project; the technical and logistical problems that it poses, including data and specimen collection and management; the ethical, legal, and social issues that it raises; the nature of the organizational structure that it requires; the relationship of this project to other endeavors, such as the Human Genome Project (HGP); and, if possible, matters of cost, duration, and management.

The committee met on 4 occasions: on April 22-23, 1996, in Washington, DC; on July 24-26, 1996, at the National Academy of Sciences Beckman Center at Irvine, California; on September 16-18, 1996, in Washington, DC; and on November 20-22, 1996, in Irvine. In addition, to elicit as wide a spectrum of opinions on the merits of the proposed survey as practical, the committee circulated a questionnaire encouraging those who could not attend the public sessions held as part of the committee's second and third meetings to submit their opinions in writing. Their comments were tabulated and taken into account in the committee's deliberations.

At its first meeting, the committee was briefed by representatives of NSF and NIH, specifically, Mary Clutter, assistant director of NSF, and Judith Greenberg, director of the Division of Genetics and Developmental Biology of the National Institute of General Medical Sciences (NIGMS). The committee then discussed its charge and how best to respond to it. It also examined its own composition to ensure that all the relevant scientific disciplines were represented and that no serious conflicts of interest existed. Because no document can be construed as the official proposal for the HGDP, the committee wrestled with the questions of what the HGDP is and who speaks on its behalf. To address the first of those questions the committee invited Luca Cavalli-Sforza, who has been an active supporter of the project, to describe the proposed survey, its aims, and the scientific issues that it addresses as he saw them. The committee then undertook the tasks of identifying persons whose experiences or known positions relative to the project were pertinent to the committee's charge, of examining the structure of the committee's final report, and of identifying subjects on which the committee would need to seek more information.

At its second and third meetings, the committee continued its fact-finding and discussions. Various spokespersons were invited to present their views of the project. Both meetings also included sessions at which members of the public and spokespersons acting on the public's behalf were invited to discuss the issues with the committee. The fourth meeting was devoted to the completion of the report that follows.

THE PROPOSED HUMAN GENOME DIVERSITY PROJECT

In its fact-finding, it became apparent to the committee that the precise nature of the proposed survey was more elusive than the committee had initially envisioned; different participants in the formulation of its consensus document had quite different perceptions of the intent of the project and even of its organizational structure. The committee reviewed the "consensus document" for the proposed Human Genome Diversity Project and was briefed by different participants involved in formulating the consensus document. It found there clearly was no sharply defined proposal that the committee could evaluate. However, because the consensus document addresses many of the issues—scientific, ethical, legal, and managerial—that any specific proposal must address, the committee sets out briefly the positions adopted in the consensus document and their rationale.

The specifics of the HGDP grew out of 3 major workshops that were convened to examine the need for a global effort to characterize human variability and to plan the enterprise. The first—supported by NSF, NIGMS, the National Center for Human Genome Research, and the Department of Energy—took place at Stanford University in July 1992 and focused on the statistical issues associated with population sampling. The second, supported by the same agencies, took place at Pennsylvania State University in November 1992 and dealt exclusively with anthropologic issues. The last workshop—supported by the Porto Conte Research and Training Foundation, the European Commission, the Soros Foundation, and the Human Genome Organization Europe, in addition to the aforementioned US agencies—was held in Porto Conte, Sardinia, in September 1993, at which time the views of some 80 participants were summarized in a consensus document that set forth the aims, rationale, value of, and guidelines for the design and scope of the proposed global survey.

Briefly put, the proposed survey focuses on variation between rather than within human populations and has 2 major aims: to investigate differences in the human genome by studying samples collected from populations that represent all the world's peoples, and to create a resource for the benefit of all humanity and for the scientific community worldwide. The proposed resource would exist as a collection of samples representing the genetic variation in human populations and as an open, long-term genetic and statistical database on variation in the human species.

The term "population" has many meanings; it is most often used to designate a body of persons (or other organisms) that have a common quality or characteristic, to designate a group of interbreeding organisms, or to designate a group of persons (or other organisms) that occupy a specific geographic locale. If language is used as the shared characteristic, over 5,000 human populations in the world have distinct properties and possibly distinct gene frequencies; these populations vary from tens of individuals to hundreds of millions. It is patently impractical to study all those populations, so some selection is inevitable. The planners of the proposed project recognize that any method of selection will be imperfect, is apt to be controversial, and could be divisive. Accordingly, they have focused their recommendations on the types of populations to be included in the survey rather than on the identification of specific human groups to be sampled. The types suggested are populations whose use can answer specific questions about the processes that have had a major influence on the composition of current ethnic groups, language groups, and cultures; populations that are anthropologically unique; populations that constitute linguistic isolates; populations that might be especially informative in identifying the genetic etiology of important diseases; and populations that are in danger of losing their identity as recognizably separate cultural, linguistic, or geographic groups of individuals.

The planners do not specify a sampling strategy. They recognize that sampling might be opportunistic, might involve a grid approach if the population is relatively large and dispersed, might entail a stratified proportional approach, or might follow some other strategy. The primary goal, however, remains the same: to obtain a sample that represents the genetic variation in the population under study. The proponents of the proposal recognize that the number of persons sampled can vary, but they suggest that a minimum of 25 samples be obtained from each population studied, although their suggested norm is 150. Sample sizes may be even larger in some instances, particularly if the aim of the research is to study phenotype-genotype associations.

The HGDP consensus document strongly recommends that all samples collected and studied be tested against a predefined core set of alleles or other genetic markers. The planners recognize, however, that a much wider variety of other markers will eventually be studied, so they require that the choice of specimens and their method of collection provide for possible expansion of tested markers. The core set has not yet been identified at the time of this study, but the planners have examined the usefulness of classical markers, restriction fragment-length polymorphisms (RFLPs), minisatellites, Y-chromosome polymorphisms, and microsatellite loci. They recommend that emphasis be placed on polymerase chain reaction (PCR)-based systems to type polymorphic markers in as much as such systems are highly flexible, are portable, are fast, are amenable to automation, and can be conducted in a multiple-locus format. But the planners note that the choice of markers should also be contingent on other considerations, including that they are technically robust, can be determined in regional and local laboratories, and ideally can be detected with nonradioactive test components.

Administratively, the implementation of the proposed project would entail the establishment and coordination of a number of regional activities and committees; some of the latter have already been constituted. However, the proposal does not make clear how centralized or dispersed the control of the database and specimen repository will be. Our committee believes that this is a potentially troublesome issue, given the number of polities and investigative groups that would be involved. Nonetheless, it clearly is important that the administrative structure be a flexible one that invites and encourages widespread participation in the project and that can adapt to new investigative groups as they develop the requisite skills and interest.

The planners have noted that the proposed survey poses ethical and legal issues; to address these, the regional committees, especially the North American one, have formulated a series of model ethical protocols for collecting DNA samples. The protocols set out guidelines for finding and approaching study populations, for achieving informed consent, for providing benefits to the sample populations (including medical services), and for maintaining privacy and confidentiality of the information that is obtained. The protocols also address such matters as the ownership and control of the samples and the patenting and commercial use of specimen-derived information. Finally, the protocols strongly recommend that investigators establish partnerships with participating populations, including their involvement in planning and conducting the research and disseminating the results.

THIS REPORT

This report examines the scientific merits and value of research on human genetic or genomic variation and the organizational, policy, and ethical issues that such research poses in a more-general context than the proposed HGDP. Chapter 2 discusses the scientific basis and usefulness of a worldwide systematic survey of human genetic variation. Chapters 3 and 4 address technical and

INTRODUCTION AND BACKGROUND *15*

logistical problems posed by such a project with respect to sampling design, obtaining and maintaining sufficient genetic material, access to research materials and information, and data handling. Chapter 5 examines ethical, legal, and social issues. Chapter 6 treats the organization of the project and research priorities.

2

Scientific and Medical Value of Research on Human Genetic Variation

No person or group can be the measure of humanity. All people—in every part of the world in all our cultural, linguistic, and biologic diversity—are equally human. Comparisons of the genetic variability between and within populations can do much to increase our understanding of the human species. As a consequence, the application of modern molecular techniques in the analysis of genetic variability of specimens collected from human populations throughout the world can contribute substantially to our knowledge of ourselves. The results of such worldwide studies should greatly enlarge our knowledge and understanding of the diversity of the human species.

Most research on human genetic variability has been done independently by different investigators. Their work is rarely comparable, however, because different sampling procedures are used. The studies also tend not to have broad geographic coverage, because investigator interests, ease of access, and financial resources drive the selection of the geographic region covered. There is need for better coordination, comprehensive geographic coverage, standardization of sampling and data-processing, and better compliance with basic standards. To address those needs, research on human genetic variability will require international cooperation and collaboration in the acquisition, examination, and sharing of the resources and knowledge essential to our collective understanding and characterization of the human species. For example, understanding human genetic variability can greatly improve our grasp of human history and migration, of the evolution of the human gene pool, and of the basic mechanisms of genetic evolution. It can also provide data relevant to biomedical application. The potential effects of research in human genetic variability are addressed in this chapter.

HUMAN HISTORY AND EVOLUTION

The technologic advances of the last decade in molecular genetics—which measures variability of the genetic material directly, not indirectly through phenotypes—promise to revolutionize the study of human genetic variation. Many long-standing questions in anthropologic genetics can now be examined with this powerful new set of scientific tools. The field has long focused on the bipartite nature of human genetic variability through studies of variation within and between populations. By examining those 2 levels, researchers have tried to explain the evolutionary processes that have given rise to the current patterns of human variation. For example, the study of the diverse population structures of a large number of human groups has clearly shown that random genetic drift—the accidents of chance that result from the small size of local groups—and the flow of genes from one population to another through migration or mate exchange are responsible for much of the patterning within human populations (Crawford 1976; Harpending and Jenkins 1975; Neel 1978; Williams-Blangero 1989).

Much has been learned from such small-scale evolutionary studies, but the usefulness of the genetic data on which they are based is limited by the small number of marker loci that were examined. Recent advances in molecular and statistical techniques might substantially improve our knowledge of the evolutionary dynamics of human populations. For instance, pairwise comparisons of sequence differences among people within a population have shown that such differences retain information regarding the demographic history of the population (Rogers and Harpending 1992; Slatkin and Hudson 1991) and can be used to make inferences regarding the forces that shaped the pattern of intragroup genetic variation. Such approaches provide a specificity and accuracy of information that has not previously been possible.

Intensive studies of intragroup variation have often been successfully undertaken by single investigators. However, past efforts to examine human genetic variability between populations on a large geographic scale have been problematic because they require many investigators and worldwide collaboration. The examination of worldwide human genetic variation focuses on between-population comparisons aimed at inferring population history or population migration patterns (Cavalli-Sforza 1994; Sokal and others 1996, 1997). For example, Cavalli-Sforza's broad-scale examination of the variation between human groups has found a major concordance between language similarities and genetic relationships. Similarly, his notions, although controversial, about the spread of agriculture by migration, rather than through the spread of customs and techniques, further illustrate the kinds of questions that could be studied in much greater detail.

In general, such endeavors have been severely compromised by the paucity of comparative information—an unfortunate byproduct of the lack of coordinated collection efforts and the inability of multiple investigators to genotype large numbers of samples for large numbers of genetic loci. Once information becomes available on a large and consistent set of genetic markers for many human populations distributed around the world, much more systematic investigations of the determinants of human variation will be possible. For example, questions regarding the timing of occurrence, the number, and the size of long-distance migration from Siberia to the Americas and the effects of such migration on the genetic variation within and between Native American populations will be easier to answer. Moreover, despite the relatively short time that has elapsed since those migrations, cultural differences between Mayans and US-situated plains Indians are striking, and it would be of interest to know whether these differences are reflected in genetic differences. Similarly, phylogenetic questions regarding events that occurred over a much-longer period could also be more-precisely answered. For example, considerable controversy exists about the contributions of early Homo erectus populations outside Africa to the genetic makeup of modern human populations. One theory is that after Homo erectus spread over most of Eurasia and Africa, geographic races developed and evolved together, with genetic interchange occurring among them as they evolved into modern humans (Wolpoff 1989). Another theory holds that anatomically modern humans emerged in Africa and spread over the Old World, replacing earlier populations without having substantial genetic interchange with them (Cann and others 1987; Stringer 1988; Stringer and Andrews 1988). Evidence from mitochondrial DNA and other genetic material used in support of the latter position is now known to be based on samples too small to test the hypothesis rigorously (Ayala 1995; Spuhler 1988; Templeton 1993). More-sophisticated sampling of more-diverse populations should help to resolve the issue.

The study of the determinants of human biologic variation has often focused on readily observable phenotypic characteristics, such as height, weight, facial dimensions, skin color, and body composition. Those traits have complex underlying genetic components, and we know little about the evolutionary processes that have led to the extensive variability in them among current human populations. Attempts to relate the regularities of observed spatial variation in such phenotypes with environmental factors have been problematic. This stems largely from the difficulty in taking into proper account the correlation between the measurements, such as height and weight or weight and blood pressure, used to characterize variation. A common goal of such studies is to test whether observed variation requires some nonrandom force, such as natural selection, so researchers must be able to predict what would be observed if variation were random. Random variation is a quantitatively definable null hypothesis from which to infer selection. With a large set of independent, neutral genetic markers, scientists can predict the effect of random variation from estimates of kinship between the populations on which phenotypic information exists. Such estimates serve as records of a population's evolutionary history. By controlling for that history, researchers can determine whether the pattern and magnitude of phenotypic variation is consistent with random evolutionary forces or whether a more-complex explanation is required. The availability of environmental measures will make possible powerful tests of the association between phenotype and environment in which the statistically confounding effects of shared genetic factors are minimized.

BASIC MECHANISMS OF GENOME EVOLUTION

Surveys of genetic variation within and among human populations can be used to address a number of basic questions in evolutionary genetics. In very few instances (in any organism) do we know the relative importance of various evolutionary forces in determining the amount and character of genetic variation within and among populations. Examples of those evolutionary forces include mutation, types of natural selection that can vary temporally and spatially, genetic drift, recombination, and migration. In particular, we are largely ignorant of the answer to a question of fundamental importance to evolutionary biologists: How much genetic variation is directly subject to natural selection—that is, how much genetic variation has direct phenotypic or functional effects that influence the survival and reproduction of individuals? Studies of genetic variation have the potential to answer such questions.

Such studies undertaken on nonhuman organisms have shown that one of the most powerful means of demonstrating and characterizing the effects of natural selection is to contrast the patterns of variation in different regions of a chromosome, at different loci, and at different sites within a gene (Aguade and Langley 1994; Akashi 1995; Begun and Aquadro 1992; Eanes and others 1996; Hudson and others 1987; Langley and others 1993; McDonald and Kreitman 1991; Sawyer and others 1987; Stephan and Mitchell 1992). Kreitman and Akashi (1995) provide an excellent review of many of the methods and their applications. Unusually high variation at a locus can result from a form of natural selection called balancing selection (Hudson and Kaplan 1988; Kreitman and Aguade 1986) that maintains polymorphisms at such loci as the HLA. Unusually low variation at a locus can be caused by a recent fixation of a mutation that is favored by selection (Kaplan and others 1989; Maynard-Smith and Haigh 1974). If a particular kind of variation (such as amino-acid changing variation) has fewer low-frequency variants than other kinds of variation, that can indicate that natural selection is acting against the low-frequency variants (Sawyer and others 1987). Contrasting the amount of variation within species with the amount of divergence between species for different classes of variation can also indicate some forms of selection (Akashi 1995; Hartl and others 1994; Sawyer and Hartl 1992). Methods that contrast different types of variation are powerful because they control for demographic and phylogenetic factors, which are necessarily the same for different loci. Studies of human genetic variation have the potential to be particularly informative with respect to those evolutionary issues.

If such studies were done on humans, they could also elucidate genomic features peculiar to our species. Many studies suggest that expansions in triplet repeats within some genes that underlie neurologic disorders (fragile X, myotonic dystrophy, Huntington disease, and so on) might be found only in human populations. Whether that type of genomic expansion is due to the unique history of the human species or to unique mutational or other genomic mechanisms is unknown. Recent studies have shown that the triplet repeats are polymorphic in all human populations tested, but populations with higher numbers of repeats (alleles) have higher incidences of the disorders. The incidence of disease can depend strongly on the biologic origins of a human population.

Medical examinations of people around the world can also bring to light new mechanisms that can be difficult to identify in other organisms. For example, the insertion of putative genetic elements known as retrotranspositions (specifically, the short interspersed repeated DNA sequence Alu and the long interspersed repeated sequence LINE1) into actively transcribed genes, such as the antihemophilic factor VIII and neurofibrin, observed as new mutations in patients, has revealed that retrotranspositions are a feature of the human genome. Because some Alu elements are polymorphic, they are useful genetic markers for the study of recent human population differentiation (Batzer and others 1996). Another recently discovered genomic feature is that some human chromosomal telomeric segments–segments at the extremities of a chromosome–show greater similarity, in sequence, to telomeric segments on nonhomologous than on homologous chromosomes. Whether that feature occurs on some or all chromosomes or in some but not other human populations is unknown. Worldwide studies of genetic variation should contribute substantially to answering this question.

BIOMEDICAL APPLICATIONS

The kinds of anthropologic, evolutionary, and migratory and other historic information to be derived from worldwide studies would involve intergroup comparisons based on sampling procedures that do not identify individual persons. They do not require intensive studies of variation within human groups, especially those using family-level differences and extensive phenotypic information, including medical, environmental, and occupational data on each individual. Epidemiologic information can be derived directly from survey data that does not identify individuals. For example, noninsulin dependent diabetes mellitus appears to be emerging as a global public health problem yet there is surprisingly little information on its frequency in many of the world's populations. The specimens from which DNA is to be derived could readily be tested for such parameters as glycosylated hemoglobin. These tests are inexpensive and do not require a high degree of technical skill, and would provide useful, albeit crude, estimates of the frequency of noninsulin dependent diabetes mellitus in all of the populations sampled. Such information could be useful to public health authorities responsible for the health of such populations and could provide the basis for more comprehensive studies of the disease's etiology.

Another example of a biomedical issue to whose understanding a large set of clearly defined population samples could contribute is linkage disequilibrium. Linkage disequilibrium refers to the statistical nonindependence of 2 gene loci. The strength of such disequilibrium is a function of population structure and history and of the physical distance between the 2 loci. Disequilibrium decays every generation, and in large populations it is found only between very tightly linked loci. However, depending on the antiquity of genetically important events in a population's history, disequilibrium can be observed in some populations to span much-larger chromosomal regions. For example, in recently admixed populations, disequilibrium can be observed over areas as large as 5-10 cm, or some 2-5% of the estimated linkage length of the average human chromosome (Morton 1991). But in small isolated populations, linkage disequilibrium might be extensive. Disequilibrium can be exploited to help map loci that are involved in disease or other complex phenotypes. Ultimately, association studies that test whether a genetic marker is statistically correlated with a disease-related phenotype can be conducted (Risch and Merikangas 1996). Because random events and the peculiarities of population structure are so important in the generation and maintenance of linkage disequilibrium, it will be useful to screen large numbers of populations for the presence and magnitude of linkage disequilibrium. This will entail selecting some loci to reflect various linkage distances, including some that are very tightly linked. Populations with more-extensive regions of linkage disequilibrium can then be identified and studied more completely for the complex inherited diseases that they experience.

As the foregoing suggests, most biomedical applications require more-specifically targeted studies involving sample sizes and sampling procedures under sampling levels 4 and 5, as described in chapter 3. Potential biomedical applications employing these more complex sampling strategies might include

- Specific gene-disease relationships (variations both within and between groups in disease susceptibility or resistance according to genotype), natural selection, and genetic drift;
- Gene-environment interactions—phenotypic variation and pharmacogenetic and toxicologic implications;
- Linkage disequilibrium associated with relatively common disorders, such as diabetes, asthma, and bipolar disease;
- Complex genetic traits associated with specific aberrant behavioral characteristics, such as aggressivity and alcoholism;
- Genetic factors linked to multivariate non-disease-related characteristics, such as height, intelligence, and aging;
- Genotype associations—predisposition to particular cancers, autoimmune disease, and other disorders;

- HLA and other immune-response-related genotypes in diverse populations—their implications for transplantation, vaccine development, and related therapies.

Unquestionably, detailed information concerning genetic diversity in widespread human populations theoretically might be applied, in various ways, to any or all of the above biomedical issues.

The specifics of how such applications could be made more scientifically rigorous pose some dilemmas in the design and conduct of a survey of genetic variation. It is possible that a few of the above applications might be addressed directly, with narrowly confined information from each individual sampled and with small to moderate samples, but most biomedical investigations will require considerably larger samples and substantially more information on each person sampled than the committee deems practical on a global scale.

Accordingly, population-based surveys of genetic variability might best be viewed not as a mechanism to answer such critical biomedical and population-genetics questions directly, but rather as a resource from which, once established, control samples could be derived for specific biomedical projects. Moreover, the availability of a survey database and sample repository could serve as points of departure for pilot studies with biomedical applications from which more-extensive and more-comprehensive investigations might be derived.

Population-based data on genomic variation could best serve as the foundation for more-expansive, conclusive research studies in the future. Such studies might be directed to subpopulations identified through pilot studies derived from the initial population-based genomic-variability survey samples. These derivative investigations would be carried out more with the populations under study and would therefore be less likely to create social and ethical dilemmas for both the investigators and the participants involved.

In summary, although biomedical applications are clearly important ultimate goals of population-based surveys of genomic variability, it appears more realistic at this stage of planning for biomedical investigations to be viewed as secondary targets. The committee appreciates that this view will be controversial and that it could have some negative consequences, such as a lesser willingness to participate in a study that has no immediate health benefits for potential subjects.

CONCLUSION

A comprehensive survey of human genetic variability both between and within populations could map such variability and place it in social and environmental context. Careful variability sampling in conjunction with the Human Genome Project could contribute fundamentally to a new era of modern molecular medicine and transform scientific understanding of human evolution and the course of human prehistory.

3

Sampling Issues

The creation of repositories of human DNA samples and corresponding databases on human genetic variation is associated with the following technical issues:

- Should sampling be population based?
- If so, what population-based sampling strategy should be used?
- What human populations should be sampled?
- How many of those populations should be sampled, how many people should be sampled from each, and how should the samples be chosen?

Regardless of sampling strategy, the introduction of any sample into a repository raises the following questions:

- In addition to the samples themselves, what information should be collected from each person in the survey?
- What types of samples should be collected (for example, blood, buccal cells, and hair roots)?
- Should transformed cell lines be created?
- How should the repositories of DNA or tissue samples be managed?

Once the samples are available for typing, still other questions arise, such as

- Which loci should be analyzed, and what types of markers should be examined?

- How should databases containing information about the samples be constructed and managed?

Finally, a central question about the use of the samples is:

- Should there be a core set of DNA markers that will be scored for all samples in the repository?

This chapter addresses the first 5 of those questions. The remainder are considered in chapter 4.

BASIC SAMPLING STRATEGIES

In most research projects, a scientist will determine the optimal sampling design from a specific narrowly defined set of hypotheses to be tested. That design will determine which people or populations are to be sampled and how large the samples should be. The main difficulty in designing a process to collect samples and develop corresponding databases for general use in research on human genome variation is that the samples and information are intended to serve as a resource for many researchers who are investigating different sets of hypotheses. The ideal sampling design would permit different researchers to extract appropriate subsamples relevant to their specific hypotheses. Although scientists have already identified some hypotheses to be tested with such resources, the samples would also be intended to be used to test others that have yet to be developed.

In general, as the number of populations or people in a sample increases, the ability to test a particular hypothesis (that is, the statistical power) is enhanced. Often, more hypotheses can be tested with larger samples than with smaller ones. For example, a sparse sample from a geographic area often cannot differentiate a single population with gene flow that is restricted by distance from fragmented populations that are separated by barriers that prevent gene flow (Templeton and Georgiadis 1996). As more populations are sampled in an area, it becomes feasible to test such hypotheses (Templeton and others 1995).

Nevertheless, it is impractical to devise a single sampling scheme that would be amenable to testing all possible hypotheses. For example, many sampling designs in medical genetics require case-control sampling (in which a subset of the total sample contains subjects who have a specific disease state and the rest of the subjects are matched to the diseased persons with respect to sex, age, or other variables), multigenerational pedigree data, or full-sib pairs. The specific requirements of such sampling cannot be anticipated until the hypothesis is well defined. Furthermore, use of a general-purpose sample-collection scheme will itself reveal gaps or weaknesses that will need to be corrected by the gathering of more samples. Hence, a sampling scheme must allow many potential hypotheses to be tested, but any recommended sampling scheme will necessarily restrict the types of hypotheses that can be tested with common sample repositories and databases.

Table 1 presents 5 sampling schemes that will be discussed in the following sections. The ones that are most restrictive—that is, permit the fewest hypotheses to be tested—are presented first. All but the first 2 (the most-restrictive alternatives) involve population-based sampling. After the sampling schemes are described, their strengths and weaknesses are compared to determine which one would be most appropriate for a coordinated global effort.

STRATEGIES THAT ARE NOT POPULATION BASED

Strategy I is the simplest sampling scheme because its sole requirement is that the sample be representative of the human species. Thus, the sample should not be derived from a single restricted group of human beings. This scheme would yield a sample that cannot be linked to specific persons, geographic areas, or populations. Each sample is identified simply as being from a human being, and no other information is obtained. This is the least-expensive type of sample to acquire, and its collection minimizes many ethical issues at both the personal and population levels (see chapter 5). Despite its simplicity, it could be used to test some important hypotheses discussed in chapter 2. In particular, it could be used to test hypotheses about human genome evolution, patterns of genetic variation relative to functional types of DNA and location in the genome, and the total amount of genetic variation found in the human species.

The other classes of hypotheses given in chapter 2 would not be amenable to testing with a sample that does not identify individuals, or geographic areas or populations, so this sampling scheme is associated with a narrow breadth of applicability.

Strategy II differs from Strategy I in that it records the geographic location of each sampling point, but the sample cannot be linked to particular persons or populations. The geographic points to be sampled could be chosen either by using a grid method or by sampling geographic areas in proportion to the density of populations in them. All the hypotheses related to genome evolution and patterns of variation that were testable with strategy I are also testable with strategy II, but in addition it is possible to test hypotheses related to the patterns of spatial variation and some hypotheses about the geographic subdivision of humans and patterns of gene flow or migration. Hence, the utility of the sampling design for testing hypotheses has been enhanced.

STRATEGIES THAT ARE POPULATION BASED

Strategy III is the first of the population-based sampling designs given in table 1. It records not only the geographic location of a sample, but also informa-

TABLE 1 Sampling Strategies

	Non-Population-Based Sampling		Population-Based Sampling		
	I Totally Anonymous	II Geographic Location	III II + Group Identification Data	IV III + Individual Phenotypic Data	V IV + Pedigrees
Testable Hypotheses	Genome evolution Patterns of variation in the genome Overall genetic variation in humans	Same as I plus: Description and determination of spatial variation (such as, variation of loci in space, migration)	Same as II plus: Patterns of migration, gene flow, and population subdivision Hypotheses from anthropology, archaeology, history, and linguistics that should affect patterns of interpopulation variation Preliminary studies on medically relevant loci Population-level medical associations	Same as III plus: Identify specific loci for possible biomedical applications Genotype interactions Within-group variation on medical and phenotypic data Associations between genes and phenotype at individual level	Same as IV plus: Detailed studies on disease-associated genes
Relative Costs	$	$$	$$$	$$$$	$$$$$

tion provided about self-reported ethnicity, primary language, sex, age, and parental birthplaces. All the hypotheses that are testable with strategies I and II can be tested with strategy III, but this strategy broadens the universe of testable hypotheses to those related to population-level relationships and differences measured primarily with data on the frequencies of alleles (alternative forms of genes at the same locus) or haplotypes (particular states of a region of DNA—if the DNA region is a coding region, haplotypes correspond to alleles), and associated and derived statistics.

With strategy III, no individual identification or other phenotypic data would be gathered, and hence the sample could not be linked to specific individuals. This restriction is compatible with testing hypotheses about the evolution of the human genome, patterns of variation in the genome, the evolution of the human species, the geographic distribution of human variation, evidence of major migrations, and patterns of gene flow and subdivision. This scheme would also allow the testing of hypotheses arising from anthropology, archaeology, history, and linguistics, such as the timing of migrations or the spread of customs influencing patterns of reproduction, that should have detectable effects on patterns of human population-genetic variation. Also included are hypotheses in genetic epidemiology that can be tested with allele-frequency or haplotype-frequency data on populations—for example, studying HLA population variation to aid in transplant matching (a benefit primarily to minority groups in developed nations), testing for systemic and infectious disease and other phenotypic associations at the population level, measuring linkage-disequilibrium patterns (nonrandom associations between allele or haplotype frequencies in a population at 2 loci or positions in the genome) as an aid to positional cloning, and determining the genomic and geographic distribution of integrated viral sequences.

The hypotheses testable with strategy III could have value for both basic and clinical research. Work on recent human evolution and on patterns of past migrations and gene flow has yielded intriguing glimpses into the origins of anatomically modern humans (Goldstein and others 1995) and the spread of agriculture (Barbujani and others 1995; Weng and Sokal 1995). Such studies have also had a major effect on related fields of research, such as paleoanthropology (Frayer and others 1993), archaeology (Ammerman and Cavalli-Sforza 1984), and linguistics (Chen and others 1995). The recent studies on human genetic variation have generated controversy, largely because the available samples from human populations are geographically inadequate to test them (Templeton 1993). Thus, there remains a great need for the type of population-level samples that would result from coordinated global sampling of human genetic variation.

The research mentioned above addresses purely scientific questions, but it has important social and political implications. The identification of a person as a member of a particular ethnic or other social group has proved to be a major factor in the behavior of that person and others—our perception of "who we are" influences how we treat others and react to them. There is still controversy about how good the available genetic data on humans are for studying recent human evolution. However, it is accepted that the major human "races" originated very recently and have undergone minimal genetic differentiation (Takahata and others 1995) or that they have exchanged genes repeatedly throughout all recent human evolutionary history, so there is only a single evolutionary lineage of humanity (Templeton 1994, 1996). Further sampling will help to determine which of those two alternatives is correct.

Other hypotheses testable with strategy III have biomedical implications at the population level. One subject of clinical relevance is the association between specific genes and human diseases, including multigenic disorders. A classic example is the population-level genetic correlation of malaria with the beta-thalassemia alleles and with the sickle-cell allele at the beta hemoglobin locus. More-recent studies have found associations of resistance to malaria with polymorphisms in the human leukocyte antigen (HLA) system and several other genes. In fact, studies elucidating the genetic factors that lead to resistance or susceptibility to infectious diseases are becoming common, as exemplified by recent work on tuberculosis and schistosomiasis.

A recent example is more enigmatic (Ansari-Lari and others 1997). An uncommon deletion of a portion of the chemokine receptor 5 gene (CCR5) is strongly associated with resistance to HIV-1 infection. In preliminary studies, this variant appears to be peculiar to populations of European ancestry and not to occur in people with ancestors in Africa, where HIV is thought to have originated. Samples from populations around the world would provide a more-definitive explanation of this finding that might have relevance to the treatment of HIV or improve the understanding of the origin and spread of this infection. Another example of the association of a marker with a systemic disease is allelic variation at the apolipoprotein E (ApoE) locus and Alzheimer's dementia. Also, interpopulation allele-frequency differences at the ApoE locus are predictive of differences in the incidence of coronary arterial disease in those populations (Lehtimaki and others 1990; Tunstall-Pedoe and others 1994).

In all such research, identifying an association or a linkage disequilibrium between a disease variant and closely linked polymorphic alleles (that is, alleles at a locus that is close to the disease locus on the same chromosome) has been crucial for identifying candidate genes for the more-common disorders that are expected to have susceptible and resistant genotypes (Brice and others 1995; Ghosh 1995). Indeed, it has recently been suggested that polymorphisms in all human genes, once recognized, can be a powerful tool for mapping the genetic components of complex diseases or traits. Studies of polymorphisms can be especially informative if conducted in recently admixed populations (that is, populations that had been genetically isolated but have recently been interbreeding) or in isolated populations, where the recent history of admixture not only induces linkage disequilibrium over large genomic segments but might also increase the incidence of disease (Stephens and others 1994).

The utility of such isolated populations is well illustrated by research on a relatively isolated population at Lake Maracaibo in Venezuela (Gusella and others 1983) that has a high incidence of Huntington disease. Studies of this population played a critical role in the ultimate cloning and identification of the gene responsible for Huntington's disease.

As those examples illustrate, population-level data on human genetic variation can have biomedical applications. However, to bring the applications to fruition often requires more-extensive sampling. For example, to identify the gene responsible for Huntington disease, it was necessary to sample the Venezuelan population and assemble extensive pedigree data based on that sampling (Gusella and others 1983). Similarly, to confirm the role of ApoE in coronary arterial disease, longitudinal studies on people were needed (Stengard and others 1995). Such detailed and specific biomedical hypotheses could not be addressed with strategy III. **From a biomedical standpoint, a large-scale, coordinated effort with strategy III would provide a resource for initial screening to identify populations that would be most promising for detailed follow-up studies.** It would then be the responsibility of individual investigators to design the sampling scheme needed to test their specific hypotheses and to resample the relevant populations.

Strategy IV, the second of the population-based strategies given in table 1, includes biomedically relevant information on individually identifiable phenotypes, particularly disease phenotypes. All the hypotheses mentioned in connection with strategy III could be tested with this scheme, but in addition one could look for genotype-disease associations instead of the much-weaker population-disease associations possible with strategy III. However, even such an enhanced data set would still be limited to disease-association studies and could not address disease causation directly. Hidden or unknown heterogeneity in the populations sampled could easily lead to false conclusions, and additional sampling (often the gathering of pedigree data) would be needed to confirm the results obtained with this strategy.

Those limitations can be avoided by going to a third level of population sampling, strategy V, the sampling of families or pedigreed persons in a population instead of persons of unknown relationship. When pedigree data are gathered with population and phenotypic data, more-definitive phenotypic studies are possible and they have enhanced power to detect markers close enough to disease loci to produce a within-family association even in the absence of a population-level association, as in the case of Huntington chorea in the Lake Maracaibo area of Venezuela (Gusella and others 1983). Moreover, when many closely linked marker loci exhibit heterozygosity, family data often allow the construction of haplotypes with more certainty. Once haplotype data exist, additional and more-powerful techniques for looking at genotype-phenotype associations can be used (Templeton and others 1987). Therefore, this form of sampling would greatly increase the biomedical utility of a human genome sample collection.

WHICH SAMPLING STRATEGY SHOULD BE USED?

The above considerations indicate that repositories of population-based samples would be much more useful than repositories of non-population-based samples for addressing major scientific and medical questions related to human genome variation. Non-population-based sampling has other weaknesses.

Sampling strategy II does not avoid ethical and legal issues. It would evoke many of the ethical considerations (discussed in chapter 5) related to population identification that apply to population-based sampling strategies. That is because many human populations have strong geographic affinities, so in practice the population source of the sample could often be inferred with little effort. In some geographic areas where self-identified human populations of diverse origins are intermixed, such as major metropolitan areas in the United States, the sample could not be linked to a specific population or populations. Thus, one would end up with a mixed sample set, with some samples having easily identified population affinities and others not. That means that the hypotheses discussed in chapter 2 that require population identification would not be consistently amenable to testing with strategy II, even though many of the ethical issues of population identification would be incurred.

Sampling strategy I, being totally unlinkable to specific persons, geographic areas, or populations, would avoid most ethical and legal issues. Existing sample collections are sufficient to test hypotheses for which strategy I would be useful—no new collections would be needed. There have been sufficient studies to show that overall levels and patterns of human genomic diversity show little between-population variation in a quantitative sense (Barbujani and others 1997), so the hypotheses to be tested under sampling strategy I do not require sampling of any new or additional populations. Moreover, similar hypotheses have been tested with Drosophila (Aquadro 1992) using samples that are smaller than some human samples already gathered. Therefore, existing sample collections are large enough to test hypotheses under sampling strategy I; no new collection would be needed.

There are advantages and disadvantages to all 3 population-based sampling strategies. Although strategy III has the least utility for biomedical hypotheses (but great utility as a tool for generating hypotheses for follow-up studies), it has the advantage of circumventing a major source of potential controversy—the inadvertent identification of a participant (see chapter 5). There is no need for individual identification when only population-level hypotheses are being tested. If databases on human genome variation do not contain such information, the possibility of revealing a specific person's identity, either deliberately or through error, will be very small. With strategies IV and V, the data being collected would constitute medical records of the persons sampled and data-management requirements for security and confidentiality would increase substantially. Moreover, there would be a risk of future revelations and adverse effects on persons and groups.

The collection of phenotypic data, as in strategies IV and V, could substantially increase the cost and time required to obtain samples, as well as the cost of data management and quality assurance. Collection of phenotypic data in the field could also greatly increase the complexity of sample collection, thereby reducing the participation of investigators who have limited resources in a coordinated sampling project. For a given amount of money and in a given period, the number of people or populations sampled would probably have to be much lower than what is possible with strategy III. Thus, although there would be a greater ability to test some biomedical hypotheses, the overall value and power of the collection for testing a wide range of hypotheses would decline because samples would be so much smaller.

Another difficulty with adding phenotypic data is the problem of deciding which phenotypes to include. The possible phenotypic measurements of biomedical relevance are virtually limitless, but if the sampling protocol is to be practical in the field, only a small number of phenotypic measurements per person could be made. Limiting the phenotypic collection to a small number of traits obviously would be useful to few researchers. Some measurements could be made on blood samples taken in the field, but many of these require either large amounts of blood or fresh, unfrozen blood, so they would be impractical for a large-scale sampling effort. **Given that only a few phenotypes could be measured, there is no logical or fair method of choosing a standard set of phenotypes. Any attempt to choose such a set is likely to incur substantial time, expense, and logistic complexity for the project as a whole and to aid only a few research programs.**

When pedigree data are added to the phenotypic data (strategy V), all the issues of cost, time, complexity, quality control, ethics, and so forth are exacerbated. Moreover, as will be discussed later, the best sampling scheme for testing population-level hypotheses is to avoid sampling of close biologic relatives; the gathering of pedigree data would inevitably increase the cost per sample, thereby almost certainly decreasing the overall sample size. That would substantially reduce the utility of the sample and the resulting data resource for addressing many of the questions discussed in chapter 2. Again, it would aid few research programs—only those addressing a very narrow range of genotypic and phenotype issues.

Sampling strategy III offers the best balance of breadth of testable hypotheses, expense, and ethical complications.

Recommendation 3.1: A coordinated global sampling effort to develop a common resource for research on human genome variation should use a population-based sampling design in which the geographic location of the sample and self-reported ethnicity, primary language, sex, age, and parental

birthplaces are recorded. **The committee notes that the inclusion of parental birthplaces with the other information identified above could, in some instances, inadvertently reveal a particular person's identity.**

HOW TO SELECT WHICH HUMAN POPULATIONS TO SAMPLE

Many of the populations to be sampled are likely to be included as a spinoff from a project proposed for other reasons. However, it is important to keep in mind that the sampling will be open-ended and cumulative and that any large-scale project should make it possible to identify populations for future sampling subject to the periodic reassessment of existing samples. In choosing populations to be sampled, it should also be kept in mind that the primary purpose is to collect a sample that reveals the extent of genetic variation in the human species as a whole. Accordingly, everyone in the world should have a finite probability of being sampled; no population or group should be excluded in advance. The sampling scheme must therefore include not only linguistically unique populations, geographically peripheral populations, and so forth, but also human populations that are large, geographically widespread, and ethnically diverse. These large, widespread populations are also critical for testing population-level hypotheses. For example, hypotheses about the effect of technologic changes on population structure and gene flow patterns might be of interest, or HLA frequencies in recently admixed populations or genetic-disease associations with systemic diseases that affect primarily populations in developed countries. Underrepresenting such populations would yield a biased sample for studying overall human evolution and restrict the overall biomedical utility of the sample.

CONSIDERATIONS IN CHOOSING SUBJECT POPULATIONS

Within-population sampling strategies for a coordinated, global sampling effort will have to take into account the unique features of the populations to be sampled; no universal within-population scheme is possible. Nevertheless, a few guiding principles should be followed. Except for the pedigree sampling scheme (strategy V), population-level hypotheses in which the primary data type is allele or haplotype frequency are most efficiently tested when the persons sampled constitute, as far as possible, a random sample from the population that they are intended to represent. For most populations, sampling should be done so as to avoid first- and second-degree relatives unless pedigree data are to be obtained. The strategy for achieving that objective would vary from population to population. For large populations in developed nations, the sampling scheme should be stratified on the basis of geography and ethnicity; larger overall samples might be required in such populations because of the stratification. In other cases, one might wish to sample by village, clan, or other entity. For relatively uniformly distributed populations, a grid approach would be appropriate. In all cases, the guiding principle should be to obtain an adequate representation of the total population and to have the persons sampled be as unrelated (biologically) to one another as possible.

NUMBER OF POPULATIONS TO SAMPLE, AND NUMBER OF PEOPLE TO SAMPLE IN A POPULATION

The number of populations sampled is often controlled by matters of convenience and opportunity. However, the number of populations needed should be considered if current collections are to be improved. This section also considers the needs for the total number of people to be sampled in populations, even though it is recognized that samples in many populations might of necessity be very small and that technical, psychologic, or other difficulties might keep people from cooperating with a sampling project.

As mentioned before, the sampling is open-ended; the number of populations to be sampled will probably grow once a coordinated study of human genetic variation is initiated. Current resources are quite limited. For example, Cavalli-Sforza and others (1994) recently compiled the most-complete set of data on variation in the human genome. Because of noncomparable sampling, different loci being investigated, and so forth, they could assemble data on only 42 populations for 43 loci (24% of the cells in the data matrix are still missing) as the current sample to represent our best estimate of human nuclear genome variation.

For other kinds of genetic variation, the situation is worse. For example, although studies on human mitochondrial DNA have attracted much attention as a tool for exploring recent human evolution, testing with rigorous statistical criteria even fundamental hypotheses regarding whether mitochondrial variation spread around the globe through recurrent gene flow or population replacement is not possible, because few populations have been sampled in a geographically accurate manner (Templeton 1993). The only mitochondrial-DNA data set that even barely satisfied the sampling requirements of recent statistical tests designed to discriminate between gene flow and population replacement is that assembled by Excoffier (1990), which includes only 18 human populations. Consequently, even obtaining samples on 100 human populations would greatly augment our ability to test hypotheses about human evolution, population structure, and genome evolution, as well as disease associations at the population level. As the number of populations increases beyond 100, additional hypotheses could be tested (for example, discriminating between isolation by distance and population fragmentation due to gene-flow barriers).

Once the populations have been chosen, it is critical to have large-enough sample sizes within the populations. As previously stated, the size of the sample needed in any given instance is determined largely by the hypothesis. For example, if the investigator is seeking to test the historical genealogy of a particular population, a sample as small as 50 could be quite useful, but it would be inadequate for characterizing genetic variability at a particular locus. In the past, human samples have been used to look for linkage disequilibrium among markers, which often is valuable in biomedical studies. However, to obtain statistically accurate estimates of linkage disequilibria often requires large samples (of at least several hundred). Another potential use of these DNA samples would be to examine different human populations for various genetic-disease alleles. Such alleles are generally rare (almost always appearing in less than 1 in 500 people, that is, an allele frequency of less than 1 in 1,000). The probability of having a rare allele(s) in a sample is roughly proportional to the frequency of the allele(s) and inversely proportional to sample size. For example, there is a 95% chance of an X-linked genetic-disease allele being in a sample of 250 males if its allele frequency is 0.012 in the population, but if the sample size is doubled to 500, there is a 95% chance of inclusion of an allele with a frequency as low as 0.006 in the population. Table 2 presents other sample sizes and their associated allele frequencies for a 95% chance of inclusion.

Another important potential biomedical application would be to examine heterogeneity in systemic diseases related to common alleles. The power for detecting statistically significant allele or haplotype frequency differences, which will often be small, also increases with sample size, as shown in table 2. Because human populations show so little overall genetic differentiation, large samples will be needed to perform such studies. For example, the e4 allele at the Apo-E locus on chromosome 19 has been shown to have a large and significant effect on the chance of death from coronary arterial disease, the largest cause of death in the developed countries (Stengard and others 1995). When published data on the frequency of the e4 allele in various countries were coupled with published incidences of death from coronary arterial disease in males by country, the re-

TABLE 2 Sampling Properties for Detecting Rare Alleles and Discriminating Between Allele-Frequency Differences in 2 Populations

Sample Size	Minimal Allele Frequency (p) of an Autosomal Locus with 95% Chance of Being in Sample	Minimal Allele Frequency ($p > 0.1$) in 1 Population Required for Discrimination from Second Population[a]
50	0.030	0.274
100	0.015	0.216
200	0.007	0.179
500	0.003	0.148
1,000	0.001	0.133

[a]These minimal allele frequencies ensure a 90% chance of being significantly different at the 5% level.

gression of death rate on allele frequency was positive and highly significant, explaining some 57% of the variance (Stengard and others 1997). That suggests that this allele increases the rate of death from coronary arterial disease in all populations, a potentially important biomedical conclusion. However, the total range of e4 allele frequencies is narrow, 0.1-0.2 in most populations (Gerdes and others 1992). To detect significant population differentiation and clinical effects over such a small frequency range requires large samples in each population. Given the overall genetic similarity of most human populations, the situation at the Apo-E locus is not likely to be exceptional.

Both rare and common alleles that predispose to systemic disease illustrate that the larger the within-population sample, the more useful the sample collection will be and the broader the variety of researchers to whom it will be useful. Much of the expense of obtaining samples is related to the logistics of going to where a population lives, so doubling sample size from 250 to 500, or even more, would often involve only a modest increase in sampling expense and effort.

Recommendation 3.2: For any given population, samples of a few hundred to several hundred persons, or even more, should be obtained whenever possible. In larger populations when the investigator deems stratified sampling to be necessary, larger overall samples would be desirable.

SUMMARY AND CONCLUSIONS

Of the various sampling strategies discussed and summarized in table 1, population-based sampling strategy III, in which only basic group-identification data are gathered, is preferred over the other strategies since neither the data nor the specimens can be linked to specific individuals.. Strategy I does not provide a rationale for global sampling, and strategy II has many of the same ethical complications as strategy III but with a substantial restriction in breadth of testable hypotheses. Strategies IV and V could greatly increase the cost, complicate sampling logistics, raise serious ethical and security concerns, and benefit only a few investigators (although the investigations that would be so benefited have the most-direct biomedical relevance). Strategy III offers the best balance of breadth of testable hypotheses, expense, and ethical complications.

4

Sample Collection and Data Management

Among the most important benefits to be derived from an international effort to study the extent of human genetic variation is the establishment of repositories of specimens and collections of data relevant to them. These repositories can fulfill a potentially important role in human genetic and anthropologic research only if the collection and management of the samples and data are well standardized and appropriate to the needs of a large body of investigators. Ideally, the specimens would be so collected and stored as to accommodate the future developments in molecular biology that can be anticipated today. It would, indeed, be tragic if after a few years it were found that the specimens that had been stored were no longer suitable for emerging research needs. This chapter reflects the committee's thoughts on how those ends can be best served generally, but detailed guidance was not part of the committee's charge. Accordingly, given the importance of this potential resource, the committee recommends that a panel be convened to provide detailed guidance before a major sampling effort and specimen collection are begun.

SOURCES OF DNA TO BE SAMPLED

Peripheral Blood

DNA can be prepared from almost all human cells or tissues. Peripheral blood can be easily and relatively painlessly obtained from an arm, a leg, or even an earlobe. Samples can be stored at room temperature for up to a week without loss of quality of the DNA. Extraction is routine; various protocols are common to most molecular-genetics laboratories. The amount of DNA that can be obtained from a 10-milliliter (10-mL) blood sample is about 500 micrograms (μg, a millionth of a gram). One genotyping assay with the polymerase chain reaction (PCR) requires 10-50 nanograms (ng, a thousandth of a microgram) of DNA, so each standard blood sample should permit 10,000-50,000 genotypings, potentially a coverage of about 1 marker per 60-300 kilobase (kb) pairs of the human genome. This marker density should be more than sufficient for most genetic analysis.

The number of possible assays can be substantially increased if care is given to the technical details of the PCR assays. For example, 10 times as many assays are possible if multiple rounds of PCR with more than a single primer pair are carried out (called hemi-nested or full-nested PCR). Such methods are routinely used for analyzing DNA samples of less than 1 ng (see, for example, Leeflang and others 1995). If 1-5 ng were used in each assay, 100,000-500,000 assays could be carried out on each 500-μg DNA sample. A 1-ng DNA sample contains about 300 copies of each single-copy gene, so both alleles at a heterozygous locus would be sufficiently represented to allow accurate genotyping of the sample (Navidi and others 1992).

Amplification of more than one marker at a time in a sample (multiplexing) is readily achieved and is routine in many laboratories. Depending on the effort devoted to working out the required conditions, the number of loci examined in one assay could increase by a factor of 5-30.

In summary, careful design of the PCR amplification strategy by using procedures that are available today could allow 500,000 assays to be carried out on 500 μg of DNA isolated from 10 mL of blood (5 loci multiplexed on each 5 ng of DNA).

Other Tissues

When blood sampling is not feasible because of cultural or technical difficulties, alternative sources of DNA can be considered. Buccal cells have been successfully used for DNA analysis in many different applications. Surface epithelial cells are collected from the side of the oral cavity with a sterilized scraper, and up to several micrograms of DNA can be extracted. DNA can also be prepared from hair follicles, but the amount of DNA recoverable is less than that with buccal sampling.

Transformed Cell Lines

Peripheral blood can also be used as a source of white blood cells for the establishment of transformed cell lines. A subset of white blood cells (known as B cells) can be transformed by Epstein-Barr virus (EBV) infection into permanent cell cultures in the laboratory. For EBV transformation of B cells, the white blood cells are separated from the red cells by a simple sedimentation procedure and mixed with an inoculum of virus preparation. The culture is then incubated at 37°C without disturbance for 2-3 wk. During this time, a subpopulation of the cells begins to proliferate, and in about 1-2 months a permanent culture is established. The transformation process is not always successful, however. Besides possible variability in the handling of blood samples and in cell-culture techniques, there is intrinsic variation in the number of B cells among people and among samples from a given person. The overall transformation success rate in a highly experienced laboratory is about 95% with 10 mL of blood. Transformed cells can be stored in liquid nitrogen and returned to culture later with rarely any difficulties. It is already possible to immortalize other cell types, such as T cells (a particular kind of white cell) from blood and fibroblasts from skin. Development of methods for establishing cell lines from buccal samples or hair roots would be of great value.

The DNA extraction protocols for transformed cells are essentially the same as those for blood samples. These cells can be readily transferred from laboratory to laboratory. DNA can then be extracted from the expanded cell cultures to provide each laboratory an essentially unlimited supply of DNA for genetic analysis.

The advantage of having cell lines is not restricted to providing an unlimited source of DNA for marker analysis. Some questions related to biomedical applications could not be addressed simply through DNA marker analyses. For example, cell lines are needed to produce large intact DNA fragments for long-range physical mapping studies and genome analyses at the chromosomal level.

Transformed cell lines are preferred for collecting samples for human genetic-variation research. That is particularly true for populations that are small and hard to sample but that might yield interesting information about human prehistory. Either because these populations are in danger of disappearing or being extensively admixed with their neighbors or because sampling them repeatedly might be both intrusive and impractical, it is important to have a sample that can support extensive work without a need for resampling. However, the cost of establishing transformed cell lines routinely can impose a substantial financial burden on the study of human genetic variation and is not recommended, except for extenuating circumstances, such as those just cited. The long-term cost for cell-line storage and maintenance is substantial. There is an urgent need for technologies for cheaper and more-reliable cell immortalization. Collections will be made in many remote areas, so methods that allow successful transformation after long storage under nonideal conditions would be particularly important. Transformation efficiency decreases greatly with time between blood collection and arrival in the laboratory. Finally, biosafety guidelines have to be strictly enforced to protect laboratory personnel, especially when they handle tissues or body fluids collected from different parts of the world.

The use of transformed cell lines, however, poses some difficulties for variation research. Using transformed cells in some cases could affect the usefulness of short-tandem-repeat (STR) analysis. STR mutation frequencies have been shown to be higher in transformed lines than in cells taken directly from the subject (Weber and Wong 1993) Transformation might also cause large DNA rearrangements in some regions of the genome. The transformation process can also select for particular subpopulations of cells that have undergone specific sequence alterations. Moreover, B-cell cultures are not suitable for answering some fundamental biomedical questions about gene expression. Gene-expression profiles can differ among different transformed cell lines, so caution is in order if they are used to study phenotypic variations at the cellular level.

Some of the advantages of the cell-line technology become less important when new DNA-amplification, cloning, and marker-analysis methods are developed. The amount of genetic information that could be obtained with one 10 mL blood sample could increase several orders of magnitude beyond the 500,000 assays that the current technology would allow. For example, the recent development of a modified method (Cheung and Nelson 1996) of whole-genome amplification might eventually be applied to the original 500-μg DNA sample, generating perhaps as much as the equivalent of 200 mg of total DNA.

CHARACTERIZATION OF GENETIC VARIATION

Early studies of human variation in the 1960s and 1970s established the importance of choosing random loci for assessing population affinities. Including only known polymorphic-marker loci in variation studies leads to difficulties. The results are biased by the failure to account for the fact that not all genomic segments are polymorphic. The rates of change are exaggerated and the known polymorphisms (first detected in northern Europeans) might not yield unbiased estimates of variation. Using a large number of known polymorphic markers might be appropriate for making relative comparisons of variation, but it is not appropriate for making absolute comparisons, which are needed for accurate assessment of what has been called the tempo and mode of evolution. Although not now practically feasible (given the large number of polymorphic markers that are available), it will be possible in a few years to assess variation in a random collection of genomic sites about whose polymorphic nature nothing is now known. This type of study is free of much of the bias that afflicts many current studies of human variation. The emergence of nucleotide sequencing as a powerful tool for genetic analysis and the development of technology that allows rapid, accurate, and inexpensive sequencing will likely be a boon to this approach. Note that if random genomic segments are selected from genes and subregions in nuclear genes, repeated DNA, mitochondrial, Y-linked, and X-linked segments, the extent and nature of genetic variation might be related to the various biologic (genetic) differences among them.

CLASSES OF DNA MARKERS

Recently developed laboratory techniques enable systematic surveys of genetic variation in a wide variety of genomic segments, including coding sequences, noncoding sequences, 5' and 3' untranslated regions, various classes of repeated DNA, and extrachromosomal mitochondrial DNA. They have led to the discovery of various classes of polymorphic markers with different mutational properties. Scientists now have a unique opportunity to design studies, rather than analyze patterns of data gathered for other purposes, that can clarify crucial features of human variation, the relationships between groups, and human microevolution.

It is commonly assumed that closely related populations will have similar numbers, types, and frequencies of alleles at any polymorphic locus, the differences increasing with the evolutionary distance between the populations. Thus, loci with a larger (rather than smaller) number of alleles and a uniform (rather than clumped) frequency distribution should be preferred, in that at such loci the probability of chance identity between populations is lower. Moreover, a large number of loci need to be investigated to reduce further the incidence of chance identity. The polymorphic alleles at a locus arise by mutation, so loci with high heterozygosity are expected, on the average, to have higher mutation rates than loci with lower heterozygosity. Because mutation will have the effect of erasing some of the historical evidence carried by polymorphic loci, loci with varied heterozygosities should be studied; different loci will illuminate different aspects (periods) of human evolution. Molecular-genetic analyses of human DNA have shown that polymorphic variation in humans arises from all the possible mutational mechanisms base substitution, insertion-deletion, and localized duplication-deletion and all of them are useful for measuring variation. Which specific type of marker is used will depend on the evolutionary questions being asked.

Recent studies of human variation have shifted to the use of molecular-DNA markers because they are highly abundant. The Human Genome Project has completed its first goal of the discovery and mapping of human polymorphisms. Over 10,000 polymorphic human loci are known and mapped, and it is expected that variation studies will largely use these markers. In addition, a number of efforts at developing new types of markers are under way. Given their numbers, one can choose sets of markers with defined characteristics, and genotypes can be assayed with one method.

Allelic variation includes base substitution and simple insertion-deletion, as well as variation arising from varying numbers of a simple-DNA sequence motif. Historically, DNA polymorphisms were detected as restriction-fragment length polymorphisms (RFLPs) by using cloned DNA probes, either specific for genes or at anonymous genomic segments. RFLPs are due to base substitution or small insertion-deletion differences that lead to the creation or loss of a restriction-enzyme recognition site. RFLPs usually consist of 2 alleles with an average heterozygosity of 25% and were discovered primarily in northern Europeans in the process of constructing a human genetic-linkage map. They generally have known map locations in the human genome and low mutation rates. In attempting to search for yet more polymorphic markers, it was recognized that loci at which alleles differed in the number of repeated (tandem) copies of a core DNA sequence (16-72 base pairs) were common in the human genome and highly polymorphic. Several hundred of these variable-number tandem-repeat (VNTR) loci have been discovered and mapped in the human genome; they tend to be in telomeric chromosomal segments. VNTR loci generally have multiple alleles with an heterozygosity exceeding 70% but also a high mutation rate. Classical RFLP and VNTR loci are assayed with the Southern blotting method, which is tedious, is time-consuming, and requires 5 µg or more of DNA per assay. Although they have been extensively used for gene-mapping studies and in forensic applications, they have seen little use in human variation studies. It is unlikely that RFLPs and VNTRs will be used as in the past, because the assay requires a greater degree of technical skill, greater access to a cloned probe, and larger quantities of DNA than other contemporary methods.

Current experience from mapping the human genome suggests that the markers with the most-desirable properties are microsatellites, also called simple sequence-length polymorphisms (SSLPs) or short tandem repeats (STRs). Allelic variation at these loci arises from differing numbers of copies of a small tandem repeat, usually dinucleotides, trinucleotides, or tetranucleotides. These markers are very abundant; they occur once every 30 kb in the human genome and over 8,000 have been genetically mapped. STRs have multiple alleles with an average heterozygosity of 70% and can be assayed with PCR and very small quantities (nanograms) of DNA. Moreover, the DNA-sequence information required to synthesize the oligonucleotide primers necessary to analyze a locus are easily obtained electronically from international genome databases. The primers themselves can be inexpensively synthesized de novo or purchased commercially. Because SSLPs are highly informative, they are useful for a variety of human evolutionary studies, but they have a higher mutation rate than single-nucleotide substitutions. Thus, although useful for some studies, they might not be desirable for all variation studies. As a consequence, there has been renewed interest in developing a large set of human biallelic markers, including RFLPs, that can be assayed with PCR. These polymorphisms, which are expected to occur in the human genome every few kilobases, are abundant and are thought to have low mutation rates. Given their expected development over the next few years as a part of the Human Genome Project, the PCR-based biallelic markers will be polymorphisms of choice. However, it must be noted that there is an inevitable tradeoff between the selection of loci that are highly informative and heterozygous (accompanied by a high mutation rate) and loci that have lower heterozygosities (accompanied by a lower mutation rate). The specific choice will depend on the nature of a given study.

Genetic variation in human populations can be studied with respect to specific genes or anonymous segments of DNA. A considerable body of literature shows that variation associated with specific genes is less than that associated with anonymous segments; but even within genes, variation can depend on whether coding sequences (and whether the first, second, or third position of a codon) or noncoding regions (5', 3' untranslated regions or introns) are studied. The extant variation within and between human populations is the outcome of both natural genetic processes in the genome and population-genetic factors that maintain them. Thus, whether genetic variation is evaluated in genes or in anonymous segments ultimately will depend on the specific questions being asked. In general, both types of variation will be studied to assess patterns of variation in gene versus nongene regions and can be used to answer questions about the importance of natural selection in the shaping of human genomic variation.

In addition to collecting, storing, and distributing DNA samples, it might be appropriate to analyze systematically all or a portion of the DNA samples at a specified set of loci. Such an analysis could provide a balanced data set appropriate for making inferences about the historical relationships of human populations. Existing variation studies have established the importance of studying mitochondrial, Y-linked, X-linked, and nuclear variation. Each has a unique set of genetic characteristics, including mode of inheritance, mutation rate in male and female germ lines, and occurrence and rate of recombination. Each will provide a different view of human variation because they will illuminate the role of different genetic processes in the generation and maintenance of variation. Several kinds of markers might be surveyed, and these are considered below.

DNA polymorphisms that are based on variation in the number of tandem repeats at a locus are detected with electrophoretic methods. These polymorphisms include microsatellites (SSLPs) and VNTRs. Many of the techniques used for studying them are used in the Human Genome Project. Oligonucleotide primers that flank a particular polymorphic region are used in PCR reactions. The PCR products can be labeled with radioactive or fluorescent tags. The sizes of the PCR products (and therefore the specific alleles) are usually determined by measuring their mobility in acrylamide gels. DNA-sequencing machines provide an automated approach to the size analysis of fluorescence-labeled PCR products. Newer technologies, such as electrophoresis in arrays of microcapillary tubes, might increase the speed of electrophoretic analysis.

DNA POLYMORPHISMS BASED ON SINGLE NUCLEOTIDE SUBSTITUTIONS

DNA polymorphisms based on simple nucleotide substitutions and insertions-deletions are easily analyzed with PCR. Once the polymorphic region is amplified, several current methods can identify the particular alleles in the product. If the alleles vary in the presence or absence of a particular restriction-enzyme site, allele detection can consist of enzyme digestion followed by electrophoresis. The alleles could also be identified without electrophoresis by using radioactivity- or fluorescence-labeled oligonucleotide hybridization probes that are specific for each allele ("dot blots"). An especially promising hybridization protocol in active development involves the so-called "DNA chip" technology. Small silicon chips containing hundreds to tens of thousands of oligonucleotide probes at known locations are constructed. Pools of PCR product from many different loci are annealed to the chip. The alleles at each locus in the sample are identified by determining the exact locations on the chip where the different products have annealed. Alleles can be identified rapidly at many loci simultaneously. Other potentially automated methods of allele typing include oligonucleotide ligation, microsequencing, and real-time quantitative allele-specific PCR (a technique known as Taqman) if high throughput at fewer loci is desired.

Allele identification by direct sequencing of PCR product has the advantage of detecting every kind of genetic variation both previously known and unknown. Short-cut methods of studying regions previously uncharacterized for polymorphisms might also prove of value. PCR products can be analyzed with several electrophoretic methods that are capable of detecting simple nucleotide substitutions and insertion-deletion differences between different samples, for example, single-strand conformation polymorphism, density-gradient gel electrophoresis, and chemical cleavage. It should be noted that unless all measures of variation are at the level of a specific nucleotide sequence, there is always a chance that some allele-typing data will be compromised. Thus, 2 allele-specific hybridization probes for a biallelic polymorphism would incorrectly determine the genotype of a sample if it contained 1 known allele and 1 unknown allele at the same nucleotide or near it because neither of the probes designed to hybridize to the known alleles would hybridize to the new allele.

SHOULD THERE BE A CORE SET OF DNA MARKERS THAT WILL BE SCORED FOR ALL SAMPLES IN THE REPOSITORY?

It has been proposed that a common "core" set of genetic markers be genotyped in each sample accepted into the repositories, either as a repository activity or by individual scientists. The main reason for supporting this core marker genotyping effort would be the uniformity of assay conditions and data interpretation. An important consequence of core genotyping would be that a balanced and well-designed data set would be made available for statistical analysis of hypotheses concerning human evolutionary history, population genetics, and genetic epidemiology. Although that is a desirable goal, a number of difficulties surround the core genotyping concept. The types (STRs and SNPs), their chromosomal origin (X, Y, autosomal, and mitochondrial), the numbers of markers used, and the number of samples analyzed will depend on the specific questions asked and will lead to a long and ever-increasing number of lists of possible core genotyping experiments. These will be difficult to identify in advance and in constant need of revision as the questions and technologies evolve.

Recent technologic advances, however, ensure that a large number of common markers will be genotyped in many people in multiple populations. Genotyping DNA samples took considerable effort in the recent past. An investigator had to make a major commitment to generating these data, and this required substantial funds. It would have been difficult to persuade any scientist not specifically interested in the genotyping results to carry out additional genotyping experiments. A variety of new high-throughput, multiplexed, inexpensive, and robust technologies are under development that could serve this additional genotyping. The use of one type of DNA "chip" in DNA typing has already been established (Ansari-Lari and others 1997; Chee and others 1996; Kozal and others 1996). Recent experiments suggest that it is possible to genotype 250 biallelic polymorphic-marker loci on a single DNA chip; it is likely that within a year chips with 2,000 such markers will be available (Wang and others 1996). The availability of such chips will make it very likely that most people will use them, rather than conventional methods and a set of markers specific to each investigator and varying widely among them. Another advantage is that in a single assay both anonymous markers and genes can be genotyped simultaneously, with large savings in labor and thus costs. As a consequence, all populations of any biologic or anthropologic interest can be genotyped with the same set of markers.

We conclude that investigator-initiated efforts will naturally result in comprehensive screening of human genome variation. This is a better and more-flexible strategy than establishing a core set of markers in the very structure of the project.

RESEARCH-MATERIALS MANAGEMENT

The specimens acquired in the course of a coordinated human genetic-variation research effort will be its most valuable and enduring resource. It can reasonably be expected that dramatic advances will occur in laboratory-analysis technologies because of current research investments in robotics, automation, and high-speed DNA-marker detection and DNA-sequencing methods. It is plausible to envision that rapid, automated determination of large portions of a person's genome might become a routine laboratory procedure early in the next century and that such future capabilities will heighten the scientific value of stored sample collections representing populations throughout the world. Special attention must therefore be given to the management of the specimens acquired as part of an organized genetic-variation program.

Three general models exist for the acquisition, processing, storage, and dissemination of biologic specimens: fully decentralized, centralized, and regional. Each has its inherent advantages and disadvantages.

The fully decentralized model represents only a minimal change from the current status of human genetic-variation research. Specimens are acquired and stored locally by independent investigators, who perform different marker assays relevant to their own research interests on the samples and might or might not allow others to have access to their research materials. Such a scheme maximizes investigator autonomy and control and has the strength that investigators might have personal knowledge about the population studied which help to inform the design of new and related research projects. The existence of many repositories and the involvement of many persons increase the likelihood of scientific innovation, and the degree of independence and autonomy afforded to investigators is an incentive to participate in a large-scale coordinated effort.

However, the fully decentralized model has inherent disadvantages. It is relatively inefficient in that it encompasses potentially large numbers of replicated laboratory and storage facilities. Quality-control procedures are difficult and expensive to implement, and availability of samples can be subject to loss of key personnel or other unpredictable events because some participating laboratories are small. The selection of samples available for sharing can also be subject to cultural biases, representing a single investigator's personal scientific agenda. In the absence of coordination between sites, a request to acquire samples with a particular set of characteristics might have to be made to many sites and investigators simultaneously.

At the opposite end of the spectrum, a global genetic-variation research program might be built around a single coordinating site with a single laboratory that received specimens submitted from around the world, and that applied a standardized set of genetic tests to all samples. This centralized model would be expected to take advantage of economies of scale in specimen-handling and would have a lower relative cost than replicating analysis and storage facilities in many sites. In addition, quality control of analyses and control of access to specimens would be confined to a single site, and the auditing of numbers of specimens sent out and to whom they are sent would be simplified.

The natural efficiencies of a centralized model are offset by many disadvantages. Primary among these would be the perception that the project was the province of a single country or a single group of investigators and the real or imagined exclusivity that might result. The institutionalization of the effort, with involvement of fewer people, might stifle innovation. A single repository would also potentially represent a single point of catastrophic failure: a natural disaster or other calamity might result in the loss of all acquired research materials.

The most versatile model for a coordinated effort funded by US agencies would involve the establishment or designation of a relatively small number of regional centers, in different parts of the United States similar to the distribution of reagents used in the human genome project. Because the scope of human genome variation research is global, it would be optimal for the US effort to cooperate and consult at the international level with respect to establishing centers in other countries. Multiple centers would have the advantage of providing backup in the storage and processing of specimens and would serve as foci for development of special expertise. By their location in diverse regions, such centers could more easily maintain awareness of local cultural, legal, and political issues while promoting the sharing of resources and technology. Moreover, the developing technologies (for example, chips) should make it possible to distribute cheaply and widely DNAs amplified from samples, thus also easing both political issues and resource requirements. At a minimum, close consultation between the United States and other countries will be necessary to ensure that common goals are achieved.

A key disadvantage of multiple centers with backup storage of specimens is that the ownership of any individual sample will be transferred to 2 or more sites; ownership of specimens has been identified as an important concern of some groups of potential research participants. Formal procedures for exchange of specimens and frequent updating of collections at multiple sites to keep them in synchrony with one another will make the regional-centers model more expensive than the centralized model.

Regardless of the model chosen for management of specimens, a number of issues will need to be addressed in the planning and execution of a globally coordinated research effort if it includes the determination of a "core" set of markers. Among these are development of standardized protocols for sample preparation, analysis, and storage and a quality-control mechanism to assess compliance with them. The raw data on allele typing will be the basis for all the scientific conclusions that follow from the human genome variation project, so it is critical that the allele-typing data be accurate. Because different laboratories might type different populations with the same genetic markers, accurate comparisons between populations depend on the accuracy of allele identification in each. Ultimately, the nomenclature for all polymorphisms will be based on the position of variable sites in the context of the complete DNA sequence of the human genome. Meanwhile, a major effort must be made to provide working definitions, and consultation between participants must be initiated. Markers that are based on differences in electrophoretic mobility might prove especially difficult, and a common set of electrophoresis molecular-weight markers should be used by all participants in the project. A set of standard control DNAs should also be used to establish the ability of any laboratory to identify accurately specific allele sizes in analogy with the proficiency testing in forensic uses of DNA. Standard control samples should also be used by all participants in the identification of simple nucleotide polymorphisms with the methods described above. And allele nomenclature must be carefully considered.

Other important considerations, regardless of whether core markers are a component of the effort, are the following:

- A resource-allocation mechanism to monitor and adjudicate requests for both renewable and nonrenewable research materials.
- A review mechanism for determining the scientific and ethical merit of requests for specimens (analogous to an institutional review board).
- A mechanism to detect and respond to unauthorized reuse of specimens for research not agreed to by subject populations.
- If individually identifiable specimens are collected, a procedure the committee does not advocate, then a mechanism must be established for recontact with and reconsent of participating groups and persons if currently unforeseen uses of specimens arise that are beyond the scope of the original informed consent.
- Enforcement of ethical protocols, especially the right of persons to withdraw their samples if the samples are personally identifiable.

DATA MANAGEMENT

The data produced by a coordinated human genome variation research effort, regardless of its scope, must be both accurate and internationally accessible to justify the investment in such an effort. The necessary data-management technologies and methods are relatively mature and economical, but the potentially sensitive nature of genetic information on persons and groups and the prospect that the data will be transported via public data networks, such as the Internet, might add requirements for information systems that support human genetic variation data management beyond the functions normally associated with collections of biologic data.

The critical issue for data management is whether a data repository will contain information that can be used to link genetic data to specific individuals. If so, such a repository becomes, in essence, a medical record and must be subject to the standards that are applied to electronic patient records, particularly standards that concern information security and privacy. Unlike conventional medical records, a person's DNA is more than a component of current health states, it can also convey information about health risks (Annas 1993a), which might affect a person's employability, insurability, and standing in the community.

A coordinated human genetic variation project would share and benefit from technologies developed for the Human Genome Project, including information technologies. Common to the 2 efforts are laboratory methods for data generation, such data items as representations of polymorphisms at particular genetic loci, DNA sequences of various lengths, and such statistical data as allele frequencies. Both are conceived as multi-investigator, geographically dispersed projects in which specimens and data are generated at many sites worldwide. Both need to acquire, store, and communicate primary observations and secondary observations, computed or inferred annotations and conclusions, maintaining an explicit labeling and separation of both types of data. Those similarities will simplify many aspects of the management of both physical and information resources, inasmuch as a coordinated human genetic variation research effort can build on the successes of existing international programs. Some forms of data errors, such as DNA-sequence errors in nonexpressed regions, are somewhat more likely to be interpreted as useful signals (that is, as genomic variation), because there are few automated error-checking rules that can be applied. As a result, standards for representing the confidence level of data items (for example, unverified, verified by independent assay, and reviewed by human experts) will need to be a component of the data-management design in a manner analogous to existing molecular-biology and genome-related databases.

Because of the sensitive nature of personally identifiable genetic data, the security of data systems and networks will need to be addressed. Genetic-variation research will occur in an environment of evolving laboratory technologies that, with increasing efficiency and speed, might be capable of uniquely identifying at least some subjects without their consent, on the basis of unique patterns in their genomes combined with ethnodemographic and detailed phenotypic correlates. Procedures to minimize such breaches should be developed.

Similarly, the correlation of ethnodemographic and other anthropologic data with molecular data will require formal and reproducible methods for representing and defining names of population groups, locales, languages, and other anthropologically important entities. In contrast with other genomic databases, geographic coordinates (latitude and longitude) will also be useful components of core human genetic-variation data.

To maximize scientific return on investment, a coordinated human genetic variation research effort will need to make progress on several unresolved issues. These include requirements related to naming systems, acceptable security, and intellectual property rights.

There are no widely accepted representation standards for machine-interpretable sociodemographic, ethnohistoric, and other anthropologic data. An internally consistent, understandable, and maintainable system for naming the peoples of the world and their self-reported social groups and languages will be at the core of scientific questions of human evolution, migration, and population structure. In this regard, a data repository will need to adopt the convention of a data dictionary that contains the explicit definitions used by the project for named groups; shorthand labels with implicit definitions will not suffice. The project should apply lessons learned in biologic naming systems; specifically, it should expect that the meaning of group names will change and that audit trails for relating prior definitions to current ones will need to be part of the data-repository design.

To the extent that a human genetic variation data repository includes information linkable to specific persons and adopts security measures to safeguard them the project will be undertaking a social experiment in determining the acceptability of strong security measures, access, and audit controls in a scientific community not attuned to these issues. Most, if not all, scientists consider themselves to be ethical professionals capable of doing the right thing without outside interference and oversight. The imposition and enforcement of security measures common to electronic medical-records systems need not be burdensome, but clearly it will introduce elements of complexity and cost for system designers, administrators, and users alike.

Intellectual-property rights are an additional unresolved issue. In a manner analogous to licensing agreements for patent rights on products derived from human genetic variation (see chapter 5), a decision will need to be made as to whether to hold copyright on the human genetic variation database. Copyright is commonly used to protect the economic value of a published work, but it can also serve as a basis to prosecute claims against misuses of or unauthorized changes in the data. If human genetic variation data are copyrighted by the organization that produces them, the question of allowing investigators or population groups to hold copyright on the use of the data provided by them will need to be addressed also.

FUNCTIONAL REQUIREMENTS FOR HUMAN GENETIC VARIATION DATA

Size of Proposed Data Sets

The scope of any proposed project has implications for the ease with which data can be stored and communicated. For example, if an initial human genetic variation database comprised the aggregated data from 500 people in each of 100 populations, there would be 50,000 unique sample records. If each of those unit records comprised basic ethnodemographic data and 100 genetic markers totaling 2,000 bytes (characters), the resulting data collection would total about 50 megabytes. Databases that size are well within the data-storage capacity of desktop microcomputers and commonly available laptop or notebook computers. An expansion of the data by another factor of 10, to 500 megabytes, would still easily fit within a single compact disk (CD), and such media might be an attractive and inexpensive means of distributing the composite data worldwide.

Dramatic advances in the speed and economy of DNA sequencing or marker identification could, however, cause data-management concerns to become the rate-limiting step in the progress of the project. If, for example, it became feasible to generate tens or hundreds of megabytes of genetic data per person economically, the submission and distribution of the data worldwide would tax the current capacity of both physical media and public data networks. However, given recent advances in storage capacity and bandwidth, if the same tempo of progress is maintained in the future, it seems unlikely that these potential limitations would be more transitory even if progress in data management were to slow somewhat.

DATA ACQUISITION METHODS

In general, there are 3 alternative data-management designs for a coordinated human genetic variation research effort, analogous to the management of specimens. The first is a global star network, where participating investigators send samples and data to a single coordinating site. The second involves the establishment of regional hierarchies; multiple coordinating sites would use formalized procedures for sharing samples and maintaining synchronized copies of data. The third is a fully decentralized specimen- and data-management scheme; data and specimens would be maintained locally and be available on demand to qualified investigators. As previously noted, of those designs, the star network might be most efficient but suffers from the risk that the project would be perceived as the province of a single country or a single group of scientists. The fully decentralized model maximizes autonomy but is inherently inefficient and makes unpredictable burdens on participating investigators.

The most reasonable scheme for data acquisition involves the recording of core identifying data for each sample locally and submission of both a specimen and accompanying information to one of several participating local, regional, or international coordination sites. **The committee strongly recommends the creation of electronic records at the point of sample acquisition.** We recognize, however, that this might not always be possible, particularly in the developing nations; in these instances, well-designed, standardized paper forms should be used for initial data capture. Transcription errors would likely be minimized by creation of an electronic record at the point of sample acquisition. Obviously, the decreasing cost and increasing presence of microcomputers and digital-data networks argue for the creation of alternative pathways of data submission to a shared resource, including magnetic or optical media (floppy disks, recordable compact disk, or digital tape), Internet file transfer protocol (ftp), and submission via interactive forms, such as those available via the World Wide Web.

QUALITY CONTROL AND ANNOTATION

Experience with multi-investigator gene sequencing and mapping projects has shown that multiple investigators working in multiple laboratories will inevitably submit data that need additional format-checking and error-checking for missing or invalid values. Maintaining consistency of naming and the addition of annotations that depict features of interest generally require a designated facility and a group of persons trained to serve as scientific curators or editors. Some errors might be evident on inspection by knowledgeable reviewers or indicated by rule-based consistency checking with computers, but it will also be necessary to verify the accuracy of submitted data by repeating the laboratory analysis on a fraction of submitted samples. A number of statistical-sampling methods have been developed for quality control of laboratories; they generally rely on independent assay, by 2 or more laboratories, of a fraction of submitted samples or the periodic determination of a reference unknown to all participating laboratories. The specific approach chosen is less important than the commitment of the participating investigators and laboratories to adhere to systematically and consistently applied methods of quality control.

Because all data collections contain errors and the minimization of those errors requires resources, acceptable error rates for specific types of data in a repository will need to be defined by participating investigators and biostatisticians. Those error rates will probably vary according to classes of hypotheses and the statistical power needed to make inferences about them. The success and credibility of the work done by multiple participating sites, investigators, and laboratories will depend on proof that reasonable quality-control standards are in place and are implemented.

COMMUNICATION VIA PUBLIC NETWORKS

A coordinated human genetic variation research effort will benefit from the recent emergence and rapid growth of the global Internet, a network of networks that provides a communication path among computers that is widely accessible at academic institutions, businesses, and residences around the world. Although the Internet is rapidly becoming ubiquitous, it is built on technical standards designed to facilitate information exchange and sharing, and it is not optimized for secure transport and exchange of data. For genetic data that cannot be linked to a person the openness and accessibility of the Internet are desirable characteristics.

For genetic data that can be linked to a person, the transport of data over public networks adds security risks that must be recognized and minimized (as discussed below). Relatively inexpensive technologies for secure communication via the Internet are being developed and tested, and they will be available for data management in a coordinated human genetic variation research initiative, if such data are included as part of the project.

ARCHIVAL STORAGE

It may be reasonably expected that there will be multiple complete copies of the collected human genome variation data in laboratories and data centers around the world. That will confer a resistance to loss of data due to a catastrophe at any site, but it means that the data resource must be designed from the start to accommodate the updating of multiple archival sites in various ways. Prominent among these will be automated, network-based update transactions, received and broadcast by data centers in a manner similar to that used by GenBank and other international genome databases. An alternative approach would be to have a single gold-standard database copy with real-time access by sites in various nations (as by a World Wide Web forms interface) for updating and editing from multiple sites. However, the uneven availability of Internet access worldwide and the sometimes-unreliable nature of international telecommunication suggest that such a mechanism would need to be supplemented by a means of data exchange that uses physical media. As noted above, it is not thought that special computer hardware or unusually large data-storage capacity will be needed for the archiving of data from human genetic variation research.

DISTRIBUTION AND ACCESS

Existing genome-related and molecular-biology databases provide a model for distribution of and access to human genome variation data. Periodic distribution of physical media, such as CD-ROM containing part or all of the data resource, will be an economical dissemination mechanism. Internet utilities, such as ftp, could be used for bulk transfers and database updating among collaborating sites. Online access for individual queries via World Wide Web forms and specialized query programs that provide similar searching and selection of records based on various patterns or attributes of specific marker loci will be valuable to the scientific community. Resources should be provided in the project for continuing software development to ensure that the investment in data acquisition and maintenance is complemented by a corresponding investment in software tools that make access to the data easy for qualified investigators. Enforcement of rigorous access controls will be an issue, as noted above, to the extent that data in a human genetic variation data repository can be linked to specific people.

SECURITY ISSUES RELATED TO HUMAN GENETIC VARIATION DATA

In a human genetic variation data resource, in which submitting investigators cannot even identify samples and records that they have submitted for purposes of linking to individually identifiable information, security risks are minimal. Where such links can be constructed or discovered, however, security takes on a much more prominent role in human genetic variation information systems design.

The Institute of Medicine has described 3 levels of data about people in its report on electronic medical records (IOM 1991). The first is "nonprivileged," which is least sensitive, not necessarily confidential, and can be accessed by or released to anyone without a subject's informed consent. A human genetic variation data repository, if it does not contain information identifiable with specific individuals, would be in this category. The second is "privileged," which includes illness-related data and, in the context of genetic variation, some types of phenotypic information. In general, in industrialized nations, privileged data are subject to government restrictions on release and informed consent and are distributed on a need-to-know basis. The highest level of restriction is on "deniable" data, which are extremely sensitive and virtually always confidential, such as records of substance abuse, mental health, HIV and AIDS, sexually transmitted diseases, and genetic characteristics with social consequences (for example, employability and insurability). Disclosure of this type of information could result in substantial harm to a person. One conundrum of the attempt to classify information as deniable for international data repositories is that the basis for what could result in substantial harm to a person might depend on local cultures and norms.

To the extent that any human genetic variation data resource contains information linkable to identifiable individuals, its design will have to anticipate security threats and types of security risks. Security threats exist for each of the 3 states of electronic information: transmission, processing, and storage. Security risks associated with each of the possible sources of threat—outsiders, negligent authorized users, and malicious authorized users—would need to be evaluated for each of those states. There are 5 basic types of security risks for personally identifiable data (Ford 1994):

- Disclosure loss of confidentiality or privacy
- Modification loss of integrity
- Fabrication loss of authenticity
- Repudiation loss of attribution
- Interruption loss of availability

A credible model for human genetic variation data management will need to address each of those types of risk if the repository contains data that can be linked to specific persons.

SUMMARY AND CONCLUSIONS

The committee believes that at a stage when genotyping technology is evolving rapidly, it would be scientifically inappropriate and premature to designate a common core set of markers that is to be genotyped in all samples. Given advances in technology, a natural outcome will be that individual investigators will perform large-scale surveys of a large number of markers to generate balanced data sets. In spite of differences among individual investigators in sampling designs due to the different hypotheses being tested, many will use common technologies that can provide uniformity in the types and numbers of markers analyzed.

With currently available laboratory and information technologies, the material- management and data-management aspects of a coordinated human genome variation research effort do not appear to constitute a serious barrier to implementation of the project. There are multiple feasible models for specimen and data management and numerous instances of international cooperation in the creation of shared repositories of biologic tissue and data. The specimens and data to be captured, analyzed, and disseminated by the project have unique aspects, which will require attention and resources, but none of them is intractable. The most-important decision about project design will be whether it will acquire specimens and data that can be linked to identifiable persons and thereby need to meet a "clinical" standard for specimen and data security and access control.

Recommendation 4.1: Blood samples collected from human populations should be converted primarily into purified DNA.

Standard protocols would allow 10,000-50,000 assays to be carried out on DNA from a single 10-mL blood sample. The number of assays could be increased by a factor of 50-300 by multiplexing and using existing technologies designed for analysis of very small DNA samples. Transformed cell lines provide an essentially inexhaustible supply of DNA and are mandatory for some kinds of studies, but they require much more funding for their creation and maintenance.

5

Human Rights and Human Genetic-Variation Research

As science advances, new insights into the methods of science emerge. In human genetic research, one important insight has been the recognition of ethical issues in the design of basic research on human genetic variation. In short, as the scientific community seeks to conduct genetic variation studies with people from an ever-wider variety of populations, it increasingly faces the challenge of respecting the rights and interests of research subjects who participate in the research both as individuals and as representatives of groups.

The research-design questions that this challenge provokes are not new to human biology or peculiar to the study of human genetic variation; they have been encountered many times in the contexts of human population genetics, biologic anthropology, and epidemiology. However, as the scope and depth of genetic variation research expand, the stakes for both individual research subjects and the groups that they represent will increase. Moreover, as representatives of groups, the individuals who provide DNA for genetic variation research are playing a role in science that our individual-oriented norms of research ethics are ill equipped to address. It will be increasingly important for new investigators to appreciate the ethical issues that they encounter and to be able to adapt how they approach such issues to the cultural circumstances in which they would like to work. The goal of this chapter is to address that need by describing the major lessons of the scientific community's experience with ethical issues and the research-design considerations that have emerged from them.

Proposals for human genome diversity research that do not adequately anticipate the issues raised by the proposers' population focus have already proved capable of generating a remarkable amount of public controversy. One prominent call to begin a systematic collection of human genetic samples for study (Cavalli-Sforza and others 1991) produced an unprecedented international reaction, including cautionary statements from UNESCO's Bioethics Committee (UNESCO 1995), the UN Commission on Human Rights (UN Commission on Human Rights 1996), the US Human Genome Project (US Congress 1993), and numerous public-advocacy organizations (Amazanga Institute and others 1996; Mead 1996; RAFI 1993). Because a final statement of the goals and methods of such a project does not exist, it is difficult to determine what concerns are justified and even harder to suggest how the scientific community might resolve them. However, it is not difficult to understand the sources of the concerns; they flow from the convergence of several sets of public experiences that all accentuate the risks posed by genetic variation research.

This suggests an approach for our analysis. Different kinds of genetic variation research will engage the concerns expressed by the public in different ways and to different degrees. Examining each of the DNA-sampling strategies from the perspectives of the controversies that serve as background to the current debate should allow us to identify the issues that are most likely to be raised by different strategies and to assess the extent of their challenges to the design and conduct of human genetic variation studies.

CONTEXT OF CONCERNS ABOUT STUDYING HUMAN GENETIC VARIATION

Human Genetics and the Misuse of Scientific Information

One of the concerns expressed by public reaction to the call for genetic diversity research is that such research could inadvertently exacerbate, rather than lessen, the habit of assigning people to socially defined ethnic categories for political and economic purposes. This habit, of course, long predates scientific thinking about human genetics. In some forms—such as racism, tribalism, and nationalism—it is likely to continue to flourish even in the absence of any additional research on human genetic variation. But the short history of the scientific study of human biology shows that where science can be interpreted to support socially defined categories, it is often used to give authority to the social policies that the categorization is designed to support (Caplan 1994; Rex and Mason 1988). That is not the intention of the scientists involved. Often, scientists sort people into socially defined groups simply for methodologic convenience, using the groups as rough markers of human biologic lineages. Sometimes, they begin with such categories to falsify them by showing that biology belies our social classification of humanity. Most contemporary proponents of human genome diversity research, in fact, use both of those contradictory rationales (HUGO 1993). Nevertheless, when the research is designed in terms of the problematic social categories, it becomes difficult for investigators to escape the accusation that they have participated in perpetuating, rather than confronting, the social problems that the categorization creates.

The danger is illustrated by an early episode in human population genetics: the 1920-1950 study of genetic variation underlying the global distribution of human blood types. That research contributed much to our scientific understanding of blood type genetics, but it was framed by many in the scientific community in terms of the taxonomy of human "races" that was influential in US and European cultures at the time. The research was understood by many to provide an objective biologic underpinning for the culture's prevalent concept of "races" and scientific support for the variety of discriminatory social policies that had been built on racial classifications (Marks 1996; Schneider 1996). As the history of blood-group genetics suggests, scientific studies that accept and use socially defined human taxonomies as biologically based can give inappropriate substance to those categories and lend credibility to the policies that they suggest (Barkan 1992).

The Rio de Janeiro Biodiversity Summit and Genetic Exploitation

A second common theme in the international reaction to calls for human genome diversity research is concern over potential commercial exploitation of the participating individuals and social groups. The concern is extrapolated from the experiences of indigenous peoples with expatriate pharmaceutical and agricultural research efforts that led to commercially profitable discoveries for the sponsors but not for the peoples whose natural resources were used. The Rio de Janeiro Earth Summit on Biodiversity of 1992 highlighted international public concern over this trend and the resulting development, in many nations, of public policies governing the ownership and control of their indigenous biologic materials (Friedlander 1996). The coincidence between the language of biodiversity in these international public-policy debates and the call for studies of human genome diversity has now provoked public concern that international efforts to study human genetic variation might result in an analogous commercial exploitation of human genetic "resources" (Friedlander 1996). This concern has been exacerbated recently by international reactions to episodes like the US government's attempt to patent a cell line from a native of Papua New Guinea (Taube 1995).

The Human Genome Project and Genetic Discrimination

The third important backdrop to the current discussion of human genomic diversity research is the international Human Genome Project itself. In its efforts to anticipate and address the ethical implications of its genetic-mapping and -sequencing work, the Human Genome Project has succeeded in raising the awareness of both the scientific community and the public of how personal genetic information can be used by social institutions against the interests of individuals and families (Juengst 1994). The Human Genome Project's documentation of the deterministic and reductionistic interpretations of personal genetic information by health professionals, insurers, employers, governments, and the public at large has, for example, already influenced the rules by which genetic research with individuals and families is conducted.

Numerous studies in which large families and linkage analysis were used to isolate and identify human genes have been conducted over the last several decades; they developed relatively seamlessly out of older traditions of Mendelian and medical family-history studies. The accelerated pace of that research and its increasing successes have resulted in increased scrutiny of the standards of practice in genetic family studies because such studies can inadvertently reveal genetic characteristics of individuals who have not given consent in the research. Past genetic family studies often recruited subjects opportunistically and dealt with issues concerning the recording of research data on nonparticipant family members and the publication of identifiable pedigrees only as they arose (Frankel and Teich 1993). Now they are required to address such concerns in advance to ensure that participation of all those affected by a study is voluntary and informed, that families are aware of the full array of possible risks, and that privacy of genetic information is protected (OPRR 1993).

The risks associated with "genetic discrimination" are real enough at the individual and family levels to justify serious consideration of the practices of medical genetics researchers (Geller and others 1996; Hudson and others 1995). Risks are likely to be even more substantial at the level of social groups. However, translating the kinds of protections that medical geneticists have adopted for individuals and family-research subjects into protections for entire social groups might require more radical changes in the traditional professional practices of biologic anthropologists and population geneticists than the ones that medical geneticists had to face.

ETHICAL CONSIDERATIONS IN THE DESIGN OF HUMAN GENETIC-VARIATION RESEARCH

The extent to which the issues raised above become challenges in research on human genetic variation will depend heavily on the goals of the research and the sampling strategy used to achieve them. However, 2 basic principles will always be relevant in research involving humans: (1) a scientifically valid research design in which the risks to human subjects are outweighed by the expected benefits is necessary and (2) for any project that involves collecting DNA samples from individual human beings (as opposed to other sources, such as anonymized blood banks), the free and informed consent of the persons from whom the DNA is collected must be obtained.

There has been some renewed interest in cultural relativism and in requiring researchers to be culturally sensitive in carrying out research in countries and communities other than their own. Such awareness is appropriate, but sensitivity to the specific practices and beliefs of a community cannot be used as a justification for violating universal human rights. These rights must be respected by all researchers, regardless of the research rules or customs in their own countries or the countries in which the research is performed. Fundamental human-rights documents that require respect for the human rights and dignity of all people include the Nuremberg Code (1947), the International Declaration of Human Rights (1948), and the International Covenant on Civil and Political Rights (1976); these documents support the "equal and inalienable rights of all members of the human family" to choose for themselves whether and how to contribute to scientific knowledge by participating in research (Steiner and Alson 1996).

Those international documents derive the rights that they enumerate from the "inherent dignity of the human person" (Preamble, ICCPR). Not only must people's dignity and welfare be respected, but so must their individual rights. Thus, Article 7 of the International Covenant on Civil and Political Rights specifically provides that "no one shall be subjected without his free consent to medical or scientific experimentation." The requirement to obtain people's informed consent to participate in research ensures that they have enough information about a given project to weigh the benefits and risks associated with becoming involved in the research before they agree to participate. It is a recognized principle of research ethics that research involving risks not be conducted on populations who will not be able to benefit from the research if it is successful (CIOMS 1993). Consent alone cannot justify research on populations that will not be able to benefit from it because such research violates basic principles of social justice and equality. Research subjects can make a gift to researchers or humanity, but the validity of such a gift in the context of studying genetic diversity, especially of isolate populations, is too problematic to provide the sole justification for the research. Nonetheless, the most important ethical question in research is always whether the research is worth doing. Only after that question and the risk-benefit question are answered favorably is it ethical to approach human subjects to solicit their participation in research.

Therefore, it is crucial to have a complete research protocol for review before the actual consent form and process for obtaining consent can be designed and evaluated. For any specific goal-oriented protocol, it should be possible to anticipate the risks and benefits to the subjects and pursue informed consent accordingly. For projects that are not able to specify goals in sufficient detail to quantify risks and benefits reasonably, the worst-case scenario should be assumed: the benefits will be at the lowest anticipated level, and the risks at the highest. That means that the burden of proof for any DNA-sampling project that does not have a well-defined hypothesis will be high. It also underlines the most basic starting point for all ethical analyses of genetic-variation research, regardless of which model is pursued: de-

fining a hypothesis and determining the benefit of knowing whether it is true.

Studies Involving Geographic, Non-Population-Sampling

The collection and storage of random samples of human DNA (sampling strategy I) that cannot be linked to identifiable persons, geographical areas, or populations would pose the fewest ethical concerns for genetic variation research. The absence of identifiers that could be used to associate specific genomic variations with specific human populations avoids the risks of inappropriately treating socially defined groups as biologic lineages and exacerbating existing social problems. By the same token, the collection of samples with this strategy need not be organized in terms of and in consultation with recognized social groups. No identifying correlations would be made, so negotiation of terms of participation with the individuals who are the sources of DNA would also be avoided because no direct benefits or risks to the individuals would flow from their participation. Finally, because no individuals would be identifiable through such a collection, protection of the rights and interests of identified individual human subjects in the later control of sample uses would be obviated.

Studies Involving Geographic Grid-Based Sampling

Sampling strategy II also avoids the need to address many of the ethical considerations discussed earlier in this chapter. The fact that the samples cannot be linked to identifiable individuals or populations allows researchers to avoid the complexity of human-subject protections and social-group interests that are involved in strategies that require more identification. However, the extent to which these issues can be avoided depends on the size of the grid used. Grids whose resolution makes it possible to isolate individual nations or populations could result in associating the geographic location of a DNA sample's source with a particular people. The ethical and social dynamics of the research would then change considerably, in ways that are best illustrated by considering the next sampling strategy, in which populations are explicitly identified.

Studies Involving Population-Identification-Based Sampling

The ethical challenges of genetic variation research increase with sampling strategy III and are exacerbated with sampling strategies IV and V. In these sampling strategies, specific human groups are identified as sample sources, and the subjects assume the role of representing the group. The social categories that define them might often be artificial from the biologist's point of view, but social groups are real human entities with both rights and interests to be respected and protected. Two issues of research design are particularly important in this respect: identifying groups to be sampled and obtaining group concurrence and involvement.

Identifying Populations To Be Sampled

Using social identities as the basis for defining populations for the study of genetic variation is probably the most controversial and problematic aspect of genetic variation research. The reasons are both scientific and political. From a scientific perspective, some genetic variation studies seek populations that function as demes: endogamous (interbreeding) populations that are substantially reproductively isolated from other populations. Intergroup comparisons and within-group comparisons typically require identifying human populations that function as demes. Many socially defined groups can satisfy that criterion from small geographically isolated communities to ethnically heterogeneous (but still largely endogamous) nations like the United States. But many cannot, and identifying them for scientific purposes can be difficult without an adequate understanding of the social and political structures of the areas being studied. Some human demes have fewer internal barriers to gene flow—created by stratification by class, caste, ethnic group, or clan affiliation—within the population than others.

The fact that a group name exists for political purposes can be scientifically misleading. Centralized authorities, whether concerned with the overall unity of a people or with the dominance of a particular group, can affect how humans are grouped despite their biologic connections. They can emphasize the homogeneity of the people that they recognize officially as groups. They can discount other claims to "peoplehood" in a country's body politic, or they can foster a particular process of national unification or self-interested differentiation in a country on the basis of hopes for the future more than present (or past) circumstances (Dominguez 1986, 1989; Gladney 1991; Handler 1988). Surveys of self-identified minority groups in given locales typically reveal greater heterogeneity than that recognized by central government authorities unless it benefits the central government to declare that a group of people claiming "peoplehood" is too heterogeneous to be considered a group for political purposes. Regional-level, such as state, governments might give people the name of the administrative center under whose jurisdiction they fall. Or people in a region might be referred to collectively by a name given to them by people with whom they trade. For example, the so-called Nakanai of New Britain were named by the people from the Rabaul area who traded with them; they now refer to themselves as Nakanai. In fact, however, the Nakanai comprise several linguistically distinct groups of villages with different histories. Individuals identify themselves to one another not by cultural or linguistic group names, but by village and clan, the socially important (and not biologically irrelevant) units in their lives. If a sample drawn from an area does not account for language groups, village locations, and the clans of the DNA sources, the information essential for both within-group and between-group comparisons will be lost.

Another challenge in identifying populations for study is that, as the case of the Nakanai illustrates, the socially defined groups that we use to identify ourselves are always internally differentiated. **Investigators will have to decide in advance what level within a given hierarchy of human organization to use to identify groups relevant to a particular study of human genetic variation and by which criteria.** Some have suggested that perhaps the fairest and most-revealing approach would be to solicit the advice of local populations to identify the level of analysis and the group identifications that would make the most sense to them. In theory, that approach could be used to protect the autonomy and interests of self-identified human groups and to avoid the use of distorting labels. However, as the Nakanai example suggests, it is not always a solution for accurately defining groups. Much will depend on from whom advice is sought on the composition and identification of human social groups. Multiple interviews should be conducted to avoid relying exclusively on 1 or 2 sources.

Accurate identification of population units for sampling purposes requires extensive knowledge of the social, political, and linguistic composition of the region to be sampled. Published ethnographic studies can provide some of this knowledge, as can anthropologists who work with the peoples. If this information is not available, researchers are advised to study the local situation in consultation with local leaders, experts, and other researchers before designing the sampling strategy.

Group Concurrence and Involvement

Many in the field believe that it is necessary to involve the social group itself in the design and implementation of a local sampling plan. This requirement challenges standard research practices at two distinct stages in the research process. The first stage, which we will call consultation, involves the initial invitation of a potentially participating group. Investigators should involve appropriate community representatives in the design of their sampling strategies, collection methods, and reciprocity agreements before any plan to sample a particular group is considered final or any individual is approached for consent. This process could take different forms with different populations: meeting first with local scientists in one context, with community leaders in another, or with lay groups devoted to particular genetic diseases in another.

The next stage of community involvement is obtaining approval of the group to participate in the study that the first stage of consultation generates. This requires activating the process that the group uses to make collective decisions on issues related to their corporate identity and interests. The process is analogous to the informed consent obtained from the individuals providing samples, and it requires the same forms of information disclosure by the investigators.

The concept of community approval is not well articulated in contemporary research policies, but it is similar to some forms of community consultation already used in some population-level research. The communication processes between the researcher and the group and the community decision-making processes that will be required to achieve group concurrence will necessarily vary from group to group. But to the extent that the research has the potential to affect the social interests of the group as a whole, any research on members of groups or communities (as defined by themselves or the researchers) must develop a protocol for community consultation and a mechanism for community input into how the research is designed, how the research itself will be conducted, and how the results will be used.

The concept of group approval has limits. Recently, the World Health Organization and the Council for International Organizations of Medical Sciences updated their "International Ethical Guidelines for Biomedical Research Involving Human Subjects" (1993). Those guidelines, which are based on the Nuremberg Code (1947), the World Medical Association's Declaration of Helsinki (1964), the Universal Declaration (1949), and the International Covenant (1976), articulate specific requirements. Even when community consent is obtained as a prerequisite for conducting research with group members, investigators must obtain the informed consent of individual prospective subjects (guideline 1). The information needed to obtain informed consent must be conveyed by the investigator in language that the subject is capable of understanding, including the research, its duration, reasonably expected benefits, foreseeable risks, alternative procedures, the extent of confidentiality, the availability of compensation, and the facts that participation is voluntary and that the subject may withdraw at any time without penalty (guideline 2). If the investigator has difficulty in communicating with the prospective subjects "to make prospective subjects sufficiently aware of the implications of participation to give adequately informed consent, the decision of each prospective subject on whether to consent should be elicited through a reliable intermediary such as a trusted community leader." However consent is obtained, all prospective subjects must be clearly informed that their participation is entirely voluntary and that they are free to refuse to participate or to cease to participate at any time without loss of any entitlement (commentary on guideline 8). In locations where women's rights to self-determination are not recognized (and thus their informed consent not possible), "women should not normally be involved in the research" (commentary on guideline 11 of the International Ethical Guidelines for Biomedical research involving human subjects), because it is likely that they will not have the freedom and power to choose whether to participate. While it is obviously wrong to exclude women from participation in a study that could lead to results from which they could benefit, it is equally important to insist on informed consent that is freely given.

Current international policy does not address whether a community should be able to veto the voluntary participation of individual members in legitimate research. If the group has decided not to participate, should individual volunteers who identify themselves as group members continue to be recruited and enrolled as representatives of the group? This conflict is particularly likely to occur in situations in which, for example, expatriate or immigrant communities of some social group's location have caused it to think about participating in research in a different way from members of their group who live elsewhere. **We think that it is too extreme a position to require both group and individual consent to DNA collection for genetic-variation research. Nonetheless, researchers will have to make sure that their participants understand both the objections of their community and the rationale for them as part of the informed-consent process and, when doing research that is opposed by a specific community, will also have to take into account the possible impact of doing such research on the likelihood that other communities will cooperate with other genetic-variation researchers in the future.**

Individually Identified DNA Sampling

Any sampling strategy that collects enough phenotypic, genealogic, or other ethnographic data to identify individual human sources of the DNA has the potential to put individual subjects and families at risk for confusion, intrafamilial disruption, stigmatization, and discrimination (Juengst 1996). That reinforces the need to have both individuals and families participate actively in the consent process and gives them rights and interests in controlling the use of their DNA samples and results that should override the claims of their communities or groups. Consequently, agreements regarding the research to be conducted on individually identified and population-representative samples should be negotiated at 3 levels: community, familial, and individual. At each successive level, the potential research participants should be afforded the right to decline or further qualify their participation in the study, within the limits of the agreements established by the larger groups that they represent. The potential for individual and familial identification has other implications that are not as important in other sampling strategies.

First, adequate confidentiality protections must be ensured. Studies that simply collect group-identification data about their sample sources cannot reasonably promise to protect the confidentiality of their findings about the group if the group identification is integral to the point of the study, and they should not attempt to do so. However, when information about individuals and families is collected, a promise of confidentiality is important to offer and honor. Steps must be taken to ensure that other individuals do not learn of the information derived from the sample if the sample can be linked to an identifiable individual. That is a basic concept in all genetics research, and it requires strict data management, oversight of data storage, rules for coding and disclosing data, rules regarding redisclosure, and basic data security and monitoring. **It is especially important to institute measures that will prevent unauthorized access to individually identifiable genetic information, so as to protect individual research participants from stigmatization and discrimination.** That will require a continuing monitoring mechanism to prevent breaches of confidentiality and to permit appropriate action to be taken against those who participate in such breaches. The degree of potential breaches of confidentiality and invasion of privacy might be so high that ethical conduct of such research will be impossible unless only DNA samples that cannot be linked to identifiable individuals are stored.

Second, studies that collect individually identifiable DNA must include mechanisms for follow-up about the results of the studies conducted on collected samples. In initial consultations with the communities to be sampled, investigators should agree on what information and follow-up services will be available to individuals or families who are found to have specific illnesses or a genetic predisposition to specific illnesses. In most cases of medically relevant genetic findings, arrangements involve a comprehensive protocol for the genetic screening and counseling of individuals, including mechanisms for dealing with findings, such as misidentified paternity, ambiguous or uncertain results, and the reproductive-risk implications of the information collected.

CONTROL

The extent of continuing involvement in the research by the group being sampled must be addressed. This includes whether groups or group spokespersons will be involved in monitoring the research conducted on DNA samples taken from their people, in granting permission for new uses of those samples if they are identifiable, in determining whether the group can withdraw from the research, and in determining how to share financial or other benefits.

With a DNA-databank research resource, the responsibility of the collection managers to monitor research conducted by external investigators is especially strong. Systems should be in place that aid the collection managers in anticipating the social consequences of particular research findings and help the public and the groups involved to prepare for those consequences. The collection, in effect, must assume the role of the genetic counselor for the participating groups, and administrators must be prepared to disclose the results of the testing in a responsible manner. To be consistent with practice in other fields of human genetics, the disclosure of the results of the research to the general public through scientific publications should be negotiated with the representatives of the donor groups (Powers 1993).

It is not ethically or legally acceptable to ask research participants to "consent" to future but yet-unknown uses of their identifiable DNA samples. Consent in such a case is a waiver of rights, and such waivers are explicitly prohibited by federal research regulations.

People have the right to withdraw their consent to research at any time, including the right to have identifiable samples destroyed or withdrawn. But how does that work on the community level? **Should the population itself be able to withdraw from the project? The answer might be that "community withdrawal" is not possible; if that is the case, it should be spelled out in both the protocol and the individual consent processes, as well as in the discussion of the protocol with community representatives. In general, consent and withdrawal are rights of individual research subjects and should not depend on the approval or disapproval of government authorities, however defined.** Some studies of American Indians have used the relevant tribal council both to give approval of proposed research and to review and have right of refusal to publish all research findings. That procedural protection might seem extreme to scientists, but such agreements are reported to have worked well in a variety of medical research projects. It is a way for genetic variation projects to respond to the legitimate interests of subjects and groups in the research.

COMMERCIALIZATION AND RECIPROCITY AGREEMENTS

Some proponents of human genetic variation research argue that it will be essential to negotiate arrangements regarding possible commercial benefits with the subject groups in advance of research, to make the participating groups "partners" with scientists in the research (North American Committee). Such a partnership implies that the subject groups will be given some role in determining the uses to which research results will be put.

Arrangements regarding financial interests in the products or outcomes of the research should be negotiated as part of the original project review and informed-consent process. In addition, a monitoring and enforcement mechanism, with representation of the affected groups, should be in place. One of the major lessons from the Rio de Janeiro Biodiversity Summit is the importance of economic and political considerations in negotiating research participation with identified human groups. That should not be surprising, inasmuch as social groups are usually created and sustained as a means of pursuing their members' economic and political interests. However, this adds a dimension to informed-consent negotiations that is foreign to most social and biomedical scientists: negotiating over what the participating group receives in return for participation.

Perhaps the most contentious issue in the short history of human genetic diversity research is the growing practice of patenting cell lines and gene sequences. Some indigenous populations are so averse to patenting that many researchers not only state that they will refuse commercial funding for genetic diversity research, but also explicitly promise not to patent or profit from any potentially profitable discoveries that might be made. Of course, researchers can speak only for themselves. As long as it is legal to patent human genes and gene sequences, others might obtain patents on them. Prohibiting the patenting of genes and gene sequences would require an international agreement binding at least all the major industrialized countries. Debate on this issue continues in Europe; only France has explicitly stated it will not permit patenting of genes and gene sequences. Nonetheless, much or most of the international controversy over collecting genes to study human genetic variation would disappear if the patenting of genes and gene sequences were outlawed.

Outlawing the patenting of human genes and gene sequences would solve one immediate problem, but it would not address the controversy over patenting human cell lines. The committee heard testimony from John Moore, whose spleen was used as a source of a cell line that was immortalized and patented without his knowledge or permission. When Moore discovered what had occurred, he sued his physician and the biotechnology companies that obtained the patent for his cell line. The California Supreme Court ultimately ruled that other people and companies could own John Moore's cell line and could patent it but that he himself could not assert an ownership interest in his own cells (Moore v. Regents 1990). The court stated that that result was necessary to protect the biotechnology industry, which might falter if individual ownership of cells (and DNA) were permitted (Annas 1993b; Knoppers and others 1996). After Moore briefed the committee, another witness, Abadio Green Stocel, from a Colombian group, said "If this can happen to a US businessman, what chance do we have?" These arguments raised a second possible approach that assumes at least some patenting will continue; the committee considered this argument in its deliberations.

A less-comprehensive, mutually agreeable strategy would be to require that all such patent applications include an agreement to share a set proportion of the resulting net proceeds or profits with the person whose body was the source of the DNA or, at that person's election, with the community of which he or she is a member. Or such "royalty" payments could be made to an international body such as UNESCO or WHO, for the benefit of the participating populations (Knoppers and others 1996).

A more-sophisticated and more-complicated approach would be to form an international organization to serve as a trustee and fundholder for all the sampled populations. Patents would be issued in the name of this trustee organization, which would license anyone who signs an agreement to share a portion of the net proceeds from products made from any patented gene, gene sequences, or cell line with the trustee organization. The trustee organization, in turn, would be required to ensure that the revenue benefited the participating populations, which would be represented in the trustee organization. Such an organization not only could ensure that financial fairness is observed in genetic diversity research, but also could develop, monitor, and enforce universal rules for protocol review and informed consent in such research.

CONCLUSION

Collecting biologic samples from specific individuals and families to extrapolate information about the social groups to which they belong is not a new scientific practice. However, as one research team put it, "The day of informal donations of DNA samples is past" (Hannig and others 1993). The confluence of several sets of ethical considerations gives that practice greater risks that human genetic variation researchers must recognize. Continued use of outmoded social categories to structure biomedical research (Osborne and Feit 1992), emerging possibilities for commercializing biomedical knowledge, and heightened awareness of the stigmatizing potential of genetic information all increase public concern about human genetic variation research. To the extent that human genetic variation research must continue to rely on socially defined human groups as surrogates for human demes (until technology to infer deme membership exists), the process of managing any coordinated effort to survey human diversity will be increasingly complex. For each socially identified set of samples, protocols for group consultation, consent, and control will have to be negotiated and balanced against the researchers' fundamental ethical obligations to protect the freedom, privacy, and welfare of the individuals involved, including the right not to participate in a study.

6

Organizational and Other Issues

The complexity of the proposed global research on the extent of human genetic variation leads to many administrative and organizational problems heretofore seldom encountered in human biologic research, or at least not encountered on a commensurate scale. The problems relevant to the choice of a sampling strategy and sample sizes have been set out in chapter 3, those related to the management of specimen and database repositories derived from international research on human genetic variation have been detailed in chapter 4, and those involving ethical issues have been described in chapter 5. Equitable solutions to those problems clearly are central to the success of any international effort to define better the extent of human genetic variation, and they are essential if funding agencies—such as the National Science Foundation and the National Institutes of Health—are to consider properly the scientific merits and limitations of this research. It is also important that these agencies recognize the humanitarian concerns voiced by people who are alarmed by the ethical and social ramifications that they believe are inherent in the research.

However, other vexing and potentially contentious administrative and organizational issues must also be addressed. They vary from the establishment of an effective, flexible, and transparent method of governance that acknowledges the international nature of the enterprise to scientifically mundane considerations, such as sharing of indirect costs when multiple sources of funds are available to support a specific endeavor or when more than one institution is involved in the research, as seems likely in many instances. The importance attached to those individual matters will vary from investigator to investigator, institution to institution, nation to nation, circumstance to circumstance, and time to time.

We are concerned here not with the particulars of those matters, but with their overall effect on the success of the research. For example, it is the committee's impression, which is based on a modest number of interviews and scrutiny of the results of a questionnaire, that the 2 most-important considerations that determine the willingness of many investigators to be involved are the degree of centralization of the global effort and the openness of, the accessibility to, and their possible involvement in whatever process of governance is established to manage the program. Younger investigators in industrialized countries and those of whatever age in developing nations express the concern that any international research effort, such as the one contemplated, will be quickly captured by established groups in the developed nations or be captive to them from the outset. They are not reassured by the experience of the Human Genome Project as it is known to them, particularly with respect to their own nations or nations in which they have been trained. They appear to fear that they will be relegated to a secondary role in the research, which will not allow them the opportunity to grow in skill or to equip their laboratories better, and that their contributions will go unrecognized and hence unrewarded. The committee cannot evaluate the merits or legitimacy of their concern, nor are we properly constituted to do so. However, we note that such perceptions, whether based on fact or not, are powerful determinants of human conduct. We recognize that any sense of exclusion of younger investigators or those from developing nations, or both, from the governance process undoubtedly would seriously compromise the contemplated research, particularly if no clear provision were made for the appeal of judgments that they see as capricious or self-serving, such as those related to access to the specimen and data repositories or the transfer of technologies.

It is of paramount importance that the process of governance be seen as inclusive, not exclusive. However, effective governance cannot be totally democratic, encompassing all potential investigators; some delegation of authority and responsibility is inevitable. It is how the latter occurs that is potentially contentious and where governance needs to be flexible and transparent.

Similarly, it is the committee's impression that much of the concern of lay groups, as presented by their spokespersons, involves the belief that they will not participate in the selection of the groups to be studied, in the design of the studies, or in possible financial and public-health benefits that stem from the projected research. In chapter 5, we have set out steps that we believe meet some, although possibly not all, of their concerns and apprehensions. However, we believe that early and continued consultation with the groups selected for study not only is mandatory, but is the only assurance of group as well as individual participation. Moreover, provision needs to be made for some tangible and immediate return to the individuals and communities as an outgrowth of their cooperation.

Some of the general administrative and organizational problems have been confronted by the Human Genome Project and in the context of the proposed Human Genome Diversity Project by the regional committees, such as the North

American one, that have been formed to stimulate regional and global interest in an international study of human genetic variation, to identify the problems posed by such research, and ultimately to monitor its conduct. The composition of these committees has been seen by some individuals and groups as at best elitist and at worst constituting a self-serving cabal. As a result, the actions of the committees have been endorsed by some but challenged by others, predominantly spokespersons for the historically disadvantaged. Nevertheless, the solutions offered by these committees to the ethical, political, and scientific issues posed by collaborative international research warrant careful consideration as possible paradigms. We believe these to constitute a conscientious effort to identify the issues and to offer possible solutions, but we note that the relevance of some of the solutions to the research currently projected, when viewed globally, might be debatable. This opinion rests largely on the geographic focus of the regional committees and the disparate levels of activity of the various committees. Some have been exceptionally active; others seem still to be in the formative stage, and the issues that they might identify as important in the successful conduct of the research in their regions are unknown to us. Nonetheless, those committees or something akin to them could play a fundamental role in the administration of the research. If the regional bodies are to serve their purposes as monitors and managers of the contemplated research, means must be found to support their activities and to permit active interaction between them. If that does not occur, the research will continue to be opportunistic and poorly integrated and will fail to serve the stated purposes of the international effort.

Support might occur in several ways. Possibly the simplest would be through the early establishment of an international mechanism of governance that is representative of all the regional groups, both scientific and lay, and the allocation of funds to this group to support regional administrative needs. We note that continued support will be required for the specimen and data repositories and that their support and administration could be a function of the international body established to govern the enterprise. The cost of these core activities is difficult to project and will depend much on the degree of centralization of effort that emerges. However, it should be noted that support must be seen as extending over the long term, possibly decades, if the goal of establishing a resource for future use is to be achieved.

We recognize that neither the National Science Foundation nor the National Institutes of Health is prepared or even able to fund a global survey like that contemplated, and we recommend that they seek advice on the role that they should play. Accordingly, these agencies should focus their financial support, at least initially, on projects originating in the US and expand their support to the international scene only after activities in the US are successfully launched. This recommendation might seem parochial to some, but the following arguments can be adduced in its support.

Some of the organizational difficulties we envisaged in a global study could be avoided or at least mitigated by initially focusing on an area of the world rich in the human resources and experience that the survey will require. There already exists in the US a formidable infrastructure: experienced investigators; well-equipped laboratories with members knowledgeable in the design and implementation of tissue repositories and in the management of large, computer-based data sets; and a means, through the existing system of institutional review boards, to protect the ethical and legal rights of individual participants and the privacy and confidentiality of the information that would be collected. The committee recognizes that a similar richness exists in Europe and encourages parallel activities there.

Second, as a result of centuries of immigration, the United States does have a diverse representation of the people of the world. Thus, a well-designed survey of human genetic variability there could shed some light on the extent of human genetic variation globally. It can be argued that migrants rarely constitute a random sample of the populations from which they came and that not all parts of the world are well-represented in the United States; this is certainly true with respect to a variety of cultural, social, and even physical measures. It is less well-established, however, whether these differences translate into a grossly biased sample of the genetic variability in their countries of origin, and this argument should not be used as a basis not to conduct a survey of human genetic variability in the United States.

Third, the establishment of an international effort will require defining the roles of interested investigators, on the one hand, and national and international agencies, on the other. Without defining such roles, any global survey would be correctly criticized for substituting a self-appointed set of administrators without official standing in any country for the recognized national and international agencies of governance, and it would be unlikely to succeed. It is not immediately obvious how the discussions of roles should be initiated. **The funding agencies, specifically the National Science Foundation and the National Institutes of Health, should initiate such discussions through their international offices.** These discussions will take time to bring to fruition; until a consensus is achieved, the US effort would be generating information of substantial moment relevant to the feasibility and urgency of an international study and would be identifying administrative barriers that would have to be surmounted.

Finally, the committee found its deliberations on the value, design, and implementation of tissue repositories, whether central or regional, constantly thwarted by the absence of information on what is actually available and accessible now. A number of repositories are of potential utility to many investigators, particularly in the US, such as the collections of the National Marrow Donor Program; the Center for Health Statistics through its periodic survey of the population of the United States; numerous multi-institutional programs, including the study of atherosclerotic risks in communities (ARIC); the Lipid Research Clinics; and the specimens assembled by individual investigators sponsored by the National Institutes of Health or the National Science Foundation. However, the information at our disposal was largely anecdotal and we could identify no complete enumeration of such collections. **The committee recommends that the National Institutes of Health or the National Science Foundation identify all such repositories, as well as the availability of the specimens to the scientific community in the US and elsewhere.**

We believe it to be inappropriate for us, as a committee, to intrude on the prerogatives of the funding agencies by suggesting a detailed mechanism for reviewing applications for support in this sphere of research. Those agencies are better informed on the options available to them than this committee and can therefore establish a more-flexible program of support and peer review. However, we do recommend that, whatever mechanism is chosen, the primary criterion for support be the scientific merit of the individual request for support. **We also suggest that in the assessment of merit the recommendations of this committee with regard to sampling strategy, sample size, and the collection of specimens and data be taken into account.**

References

Aguade M, Langley CH. 1994. Polymorphism and divergence in regions of low recombination in Drosophila. In: Golding GB, editor. Non-neutral evolution: theories and molecular data. New York: Chapman & Hall. p 67–76.

Akashi H. 1995. Inferring weak selection patterns of polymorphism and divergence at "silent" sites in Drosophila DNA. Genetics 139: 1067-1076.

Amazanga Institute and others. 1996. Declaration of indigenous peoples of the western hemisphere regarding the human genome diversity project. Cultural Survival 20: listed in Mead ATP. Genealogy, sacredness and the commodities market: 15 statements by regional indigenous people's organizations. Cultural Survival 20 (Summer):50.

Ammerman AJ, Cavalli-Sforza LL. 1984. The Neolithic transition and the genetics of populations in Europe. Princeton, NJ: Princeton University Press.

Anderson B. 1983. Imagined communities: reflections on the origin and spread of nationalism. London: Verso.

Annas GJ. 1993a. Privacy rules for DNA databanks. JAMA 27: 2346–50.

Annas GJ. 1993b. Standard of care: the law of American bioethics. New York: Oxford University Press. p 167–77.

Ansari-Lari MA, Liu XM, Metzker ML, Rut AR, Gibbs RA. 1997. The extent of genetic variation in the CCR5 gene. Nat Genet 16(3): 221–22.

Aquadro CF. 1992. Why is the genome variable? Insights from Drosophila. Trends in Genetics 8: 355–62.

[ASHG] American Society of Human Genetics. 1996. ASHG report: statement on informed consent for genetic research. Human Genet 59: 471–74.

Ayala FJ. 1995. The myth of Eve: molecular biology and human origins. Science 270: 1928–36.

Barbujani G, Magagni A, Minch E, Cavalli-Sforza LL. 1997. An apportionment of human DNA diversity. Proc Natl Acad Sci 94: 4516–19.

Barbujani G, Sokal RR, Oden NL. 1995. Indo-European origins: a computer simulation test of five hypotheses. Am J Phys Anthropol 96: 109–32.

REFERENCES

Barkan E. 1992. The retreat of scientific racism: changing concepts of race in Britain and the United States between the World Wars. Cambridge, England: Cambridge University Press.

Batzer MA, Arcot SS, Phinney JW, Alegria-Hartman M, Kass DH, Milligan SM, Kimpton C, Gill P, Hochmeister M, Ioannou AP, Herrera RJ, Boudreau DA, Scheer WD, Keats BJ, Deininger PL, Stoneking M. 1996. Genetic variation of recent Alu insertions in human populations. J Mol Evol 42:22–29.

Begun DJ, Aquadro CF. 1992. Levels of naturally occurring DNA polymorphism correlate with recombination rates in D. melanogaster. Nature 356: 519–20.

Bodmer J and others. 1993. Human Genome Diversity Project: summary document. London, England: HUGO Europe.

Boyce AJ, Harding RM, Martinson JJ. 1995. Population genetics of the g-globin complex in Oceania. In: Boyce AJ, Reynolds V, editors. Human populations: variation and adaptation. Oxford, England: Oxford University Press.

Brice A, Tardieu S, Campion D, Leguern E, Martinez M, Carpentier A, Penet C, Dubois B, Bellis M, Mallet J, Hannequin D, Clergetdarpoux F, Agid Y, Michon A, Pillon B, Babron MC, Calanda A, Puel M, Ledoze F, Pasquier F, Zimmermann MA, Desi M, Verceletto M, Thomasanterion C, Lemaitre MH, and others. 1995. Allelic association at the D14S43 locus in early onset Alzheimer's disease. Am J Med Genet 60: 91–93.

Cann RL, Stoneking M, Wilson AC. 1987. Mitochondrial DNA and human evolution. Nature 325: 31–36.

Caplan A. 1994. Handle with care: race, class and genetics. In: Murphy TF, Lappe M, editors. Justice and the Human Genome Project. Berkeley: University of California Press. p 30–45.

Cavalli-Sforza LL, Cavalli–Sforza F. 1995. The great human diaspora: the history of variation and evolution. Reading, Mass.: Addison-Wesley.

Cavalli-Sforza LL, Menozzi P, Piazza A. 1994. History and geography of human genes. Princeton, N.J.: Princeton University Press.

Cavalli-Sforza LL, Wilson AC, Cantor CR, Cook-Deegan RM, King MC. 1991. Call for a worldwide survey of human genetic diversity: a vanishing opportunity for the human genome project. Genomic 11: 490-91.

Charlesworth C, Morgan MT, Charlesworth D. 1993. The effect of deleterious mutations on neutral molecular variation. Genetics 134: 1289-1303.

Chee M, Yang R, Hubbell E, Berno A, Huang XC, Stern D, Winkler J, Lockhart DJ, Morris MS, Fodor SPA. 1996. Accessing genetic information with high-density DNA arrays. Science 274:610-14.

Chen J, Sokal RR, Ruhlen M. 1995. Worldwide analysis of genetic and linguistic relationships of human populations. Hum Biol 67: 595–612.

Cheung VG, Nelson SF. 1996. Whole genome amplification using a degenerate oligonucleotide primer allows hundreds of genotypes to be performed on less-than one nanogram of genomic DNA. Proc Natl Acad Sci 25:14676-79.

[CIOMS] Council for International Organizations of Medical Sciences. 1993 International ethical guidelines for biomedical research involving human subjects. Available from: CIOMS. p 1-63.

Clayton-Wright E, Steinbuerg KK, Khoury MJ, Thomson E, Andrews L, Kahn MJ, Kopelman LM, Weiss JO. 1995. Informed consent for genetic research on stored tissue samples. JAMA 274: 1786–92.

Crawford M, editor. 1976. The Tlaxcaltecans: Prehistory, demography, morphology and genetics. Anthropology 7. Lawrence, Kans.: University of Kansas Press.

Darwin CR. 1859. On the origin of species by means of natural selection, or the preservation of favoured races in the struggle for life. London, England: Murray.

Dominguez V. 1986. White by definition: social classification in creole Louisiana. New Brunswick, N.J.: Rutgers University Press.

Dominguez V. 1989. People as subject, people as object: selfhood and peoplehood in contempoary Israel. Madison, Wis.: University of Wisconsin Press.
Eanes WF, Kirchner M, Yoon Y, Biermann CH, Wang IN, McCartney MA, Verrelli BC. 1996. Historical selection, amino acid polymorphism and lineage-specific divergence at the G6PD locus in Drosophila melanogaster and D-simulans. Genetics 144: 1027-41.
Excoffier L. 1990. Evolution of human mitochondrial DNA: evidence for departure from a pure neutral model of population at equilibrium. J Mol Evol 30: 125-39.
Excoffier L, Smouse PE. 1994. Using allele frequencies and geographic subdivision to reconstruct gene trees within a species-molecular variance parsimony. Genetics 136: 343–59.
Foley R. 1987. Hominid species and stone-tool assemblages: how they are related. Antiquity 61: 380–92.
Ford W. 1994. Computer communications security: principles, standard protocols and techniques. Englewood Cliffs, N.J.: Prentice-Hall.
Fox R. 1990. Nationalist ideologies and the production of national cultures. Washington, D.C.: American Ethnological Society Monograph Series.
Frankel M, Teich E., editors. 1993. Ethical and legal issues in pedigree research. Washington, D.C.: AAAS.
Frayer DW, Wolpoff MH, Thorne AG, Smith FH, Pope GG. 1993. Theories of modern human origins: the paleontological test. Amer Anthrop 95: 14–50.
Friedlander J. 1996. Genes, people and property. Cultural Survival Quarterly 20 (Summer): 22–24.
Geller L, Alper J, Billings P, Barash C, Beckwith J, Natowicz M. 1996. Individual, family and societal dimensions of genetic discrimination: a case study analysis. Sci Eng Ethics 29: 71–88.
Gerdes LU, Klausen IC, Sihm I, Faergeman O. 1992. Apolipoprotein E polymorphism in a Danish population compared to findings in 45 other study populations around the world. Genet Epidemiol 9: 155–67.
Ghosh S. 1995. Probability and complex disease genes. Nat Genet 9: 223–24.
Gladney DC. 1991. Muslim Chinese: ethnic nationalism in the People's Republic of China. Cambridge, Mass.: Harvard University Press.
Gillis AM. 1994. Getting a picture of human variation. BioScience 44: 8–11.
Goldstein DB, Linares AR, Cavalli-Sforza LL, Feldman MW. 1995. Genetic absolute dating based on microsatellites and the origin of modern humans. Proc Natl Acad Sci 92: 6723–27.
Gusella JF, Wexler NS, Conneally PM, Naylor SL, Anderson MA, Tanzi RE, Watkins PC, Ottina K, Wallace MR, Sakaguchi AY, Young AB, Shoulson I, Bonilla E, Martin JB. 1983. A polymorphic DNA marker genetically linked to Huntington's disease. Nature 306: 324–38.
Handler R. 1988. Nationalism and the politic of culture in Quebec. Madison, Wis.: University of Wisconsin Press.
Hannig V, Clayton E, Edwards K. 1993. Whose DNA is it, anyway? Relationships between families and researchers. Genetics 47: 257–60.
Harpending HC, Jenkins T, 1975. !Kung population structure. In: Crow JF, Denniston C, editors. Genetic Distance. New York: Plenum Press.
Harper P. 1993. Research samples from families with genetic diseases: a proposed code of conduct. British Med 306: 1391–94.
Hartl DL, Mariyama EN, Sawyer S. 1994. Selection intensity for codon bias. Genetics 138: 227-234.
Hirschfeld L, Hirschfeld H. 1919. Serological differences between the blood of different races: the results of researches on the Macedonian front. Lancet II: 675–79.
Houghton P. 1996. People of the great ocean: aspects of human biology of the early Pacific. Cambridge, England: Cambridge University Press.
[HUGO] Human Genome Organization. 1993. The Human Genome Diversity (HGD) Project: summary document. Available from HUGO Europe, One Park Square West, London, England. p 1–8.

REFERENCES

Hudson RR, Kaplan NL. 1988. The coalescent process in models with selection and recombination. Genetics 120:831-40.

Hudson RR, Kreitman M, Aguade M. 1987. A test of neutral molecular evolution based on nucleotide data. Genetics 166: 153-159.

Hudson K, Rothenberg K, Andrews L, Kahn M, Collins F. 1995. Genetic discrimination and health insurance: an urgent need for reform. Science 270: 391–93.

[IOM] Institute of Medicine, Dick RS, Steen EB, editors. 1991. The computer-based patient record: an essential technology for health care. Washington, D.C.: National Academy Press.

Juengst E. 1994. Human genome research and the public interest: progress notes from an American science policy experiment. Hum Genet 54: 121–28.

Juengst E. 1996. Respecting human subjects in genome research: a preliminary policy agenda. In: Vanderpool H, editor. The ethics of research involving human subjects: facing the 21st century. Frederick, Md.: University Publishing Group. p 401–29.

Kaplan NL, Hudson RR, Langley CH. 1989. The "hitchhiking effect" revisited. Genetics 123: 887–99.

Knoppers BM, Hirtle M, Lormeau S. 1996. Ethical issues in international collaborative research on the human genome: the HGP and the HGDP. Genomic 34: 271–82.

Kozal MJ, Shah N, Shen N, Yang R, Fucini R, Merigan TC, Richman DD, Morris D, Hubbell E, Chee M, Gingeras TR. 1996. Extensive polymorphisms observed in HIV-1 clade B protease gene using high-density oligonucleotide arrays. Nat Med. 2: 753-59.

Kreitman M, Aguade M. 1986. Excess polymorphism at the Adh locus in Drosophila melanogaster. Genetics 114: 93–110.

Kreitman M, Akashi H. 1995. Molecular evidence for natural selection. Annu Rev Ecol Syst 26: 403–22.

Langley CH, MacDonald J, Miyashita N, Aguade M. 1993. Lack of correlation between interspecific divergence and intraspecific polymorphism at the supressor of forked region in Drosophila melanogaster and Drosophila simulans. Proc Natl Acad Sci 90: 1800–03.

Leeflang EP, Hubert R, Schmitt K, Zhang L, Arnheim N. 1994. Single Serum Typing. In Dracopoli NC, Haines J, Korf BR, Morton C, Seidman CE, Seidman JG, Moir DT et al. Current protocols in human genetics: Supplement 3, Unit 1.6. New York: John Wiley and Sons.

Lehtimaki T, Moilanen T, Viikari J, Akerblom HK, Ehnholm C, Ronnemaa T, Marniemi J, Dahlen G, Nikkari T. 1990. Apolipoprotein E phenotypes in Finnish youths: a cross sectional and 6-year follow–up study. J Lipid Res 31: 487–95.

Locke M. 1994. Interrogating the Human Genome Variation Project. Soc Sci Med 39: 603–6.

Marks J. 1996. The legacy of serological studies in American physical anthropology. History and Philosophy of the Life Sciences. 18:75-91.

Maynard-Smith J, Haigh J. 1974. The hitchhiking effect of a favorable gene. Genet Res 23: 23–35.

McDonald JH, Kreitman M. 1991. Adaptive protein evolution at the Adh locus in Drosophila. Nature 351: 652–54.

Mead ATP. 1996. Genealogy, sacredness and the commodities market. Cultural Survival Q 20.

Montagu A. 1964. Man's most dangerous myth: the fallacy of race, 4th ed. Cleveland: World Pub. Co.

Moore v. Regents of the University of California 793 P.2d 479, 271 Cal Rptr 146. (1990).

Morton NE. 1991. Parameter of the human genome. Proc Natl Acad Sci 88: 7474–76.

Mourant AE, Kopec AC, Domaniewska–Sobczak K. 1976. The distribution of the human blood groups and other polymorphisms, 2nd ed. Oxford, England: Oxford University Press.

Navidi W, Arnheim N, Waterman M. 1992. A multiple tubes approach for accurate genotyping of very small DNA samples using PCR: statistical considerations. Am J Human Genet 50: 347–59.

Neel JV. 1978. The population structure of an Amerindian tribe, the Yanomama. Ann Rev Genet 12: 365–413.

North American Committee of the Human Genome Diversity Project. ca. 1993. Model ethical protocol. Available from the Morrison Institute for Population and Resource Studies, Stanford University, Stanford, Calif.

[OPRR] Office of Protection from Research Risk, Department of Human Health Services. 1993. Human genetic research. In: Protecting human research subjects: institutional review board guidebook, Chapter H. Bethesda, Md.: OPRR.

Osborne N, Feit M. 1992. The use of race in medical research. JAMA 267: 275–79.

Phillips J and others. 1995. Policy statement on storage and use of genetic materials. Human Genet 57: 1400–500.

Powers M. 1993. Publication-related risks to privacy: ethical implications of pedigree studies. IRB: A review of human subjects research 15: 17–22.

[RAFI] Rural Advancement Foundation International. 1993. Patents, indigenous peoples, and human genetic diversity. RAFI Communique: May.

Rex J, Mason D, editors. 1988. Theories of race and ethnic relations. Cambridge, England: Cambridge University Press.

Risch N, Merikangas K. 1996. The future of genetic studies of complex human diseases. Science 273: 1516–17.

Roediger DR. 1991. The wages of whiteness: race and the making of the american working class. London: Verso.

Rogers AR, Harpending HC. 1992. Population growth makes waves in the distribution of pairwise genetic differences. Molec Biol Evol 9: 552–69.

Roychoudury AK, Nei M. 1988. Human polymorphic genes: world distribution. New York: Oxford University Press.

Sapolsky R, Spencer J, Rioux J, Kruglyak L, Hubbell E, Ghandour G, Hawkins T, Hudson T, Lipshutz R, Lander E. 1996. Towards a third generation genetic map of the human genome based on bi-allelic polymorphisms. Amer J Hum Genet 59:A3

Sawyer SA, Dykhuizen DE, Hartl DL. 1987 Confidence intervalfor the number of selectively neutrl amino acid polymorphisms. Proc Natl Acad Sci 84: 6225–28.

Sawyer SA, Hartl DL. 1992. Population genetics of polymorphism and divergence. Genetics 132: 1161–76.

Schneider WH. 1996. The history of research on blood group genetics: intial discovery and diffusion. History and Philosophy of the Life Sciences. 18:7–33.

Slatkin M, Hudson RR. 1991. Pairwise comparisons of mitochondrial DNA sequences in stable and exponentially growing populations. Genetics 129: 555–62.

Sokal RR, Oden NL, Rosenbureg MS, DiGiovanni D. 1997. The patterns of population movements in Europe and some of their genetics consequences. Amer J Hum Biol 9: 391–404.

Sokal RR, Oden NL, Walker J,. DiGiovanni D, Thomson BA. 1996. Historical population movements in Europe influence genetic relationships in modern samples. Hum Biol 68: 873–98.

Spuhler JN. 1988. Evolution of mitochondrial DNA in monkeys, apes, and humans. Yearbook of Anthropology: Yearbook Series Volume Anthropol 96: 183–84.

Steiner H, Alson P, editors. 1996. International human rights in context: law, politics and morals. Oxford, England: Oxford University Press.

Stengard JH, Pekkanen J, Sulkava R, Ehnholm C, Erkinjuntti T, Nissinen A. 1995. Apolipoprotein E polymorphisim, Alzheimer's disease and vascular dementia among elderly Finnish men. Acta Neurol Scand 92: 297–98.

Stengard JH, Weiss KM, Sing CF. 1997. An ecological study of association between coronary heart disease death rates in men and the frequencies of common allelic variations in the gene coding for apolipoprotein E. Circulation submitted.

Stephan W and Mitchell SJ. 1992. reduced levels of DNA polymorphism and fixed between-population differences in the centromeric region of Drosophila ananassae. Genetics 132: 1039–45.

Stephens JC, Briscoe D, O'Brien SJ. 1994. Mapping by admixture linkage disequilibrium in human populations: limits and guidelines. Amer J Hum Genet 55: 809-24.

Stringer C, Gamble C. 1993. In search of the Neanderthals: solving the puzzle of human origins. London, England: Thames and Hudson.
Stringer C. 1988. The dates of Eden. Nature 331: 565–66.
Stringer CB, Andrews P. 1988. Genetic and fossil evidence for the origin of modern humans. Science 240: 781–4.
Takahata N, Satta Y, Klein J. 1995. Divergence time and population size in the lineage leading to modern humans. Theoret Pop Biol 48: 198–221.
Taube G. 1995. Scientists attacked for "patenting" Pacific tribe. Science 270: 1112.
Templeton AR. 1993. The "Eve" hypothesis: a genetic critique and reanalysis. Amer Anthropol 95: 51–72.
Templeton AR. 1994. The role of molecular genetics in speciation studies. In: Schierwater B, Streit B, Wagner GP, DeSalle R, editors. Molecular ecology and evolution: approaches and applications. Basel, Switzerland: Birkhauser-Verlag. p 455–77.
Templeton AR. 1996. Gene lineages and human evolution. Science 272: 1363.
Templeton AR, Boerwinkle E, Sing CF. 1987. A cladistic analysis of phenotypic associations with haplotypes inferred from restriction endonuclease mapping. I. Basic theory and an analysis of Alchohol dehydrogenase activity in Drosophila. Genetics 117: 343–51.
Templeton AR, Georgidis NJ. 1996. A Landscape approach to conservation genetics: conserving evolutionary processes in the African Bovidae. In: Avise JC, Hamrick JL, editors. Conservation genetics: case histories from nature. New York: Chapman and Hall. p. 398-430.
Templeton AR, Routman E, Phillips C. 1995. Separating population structure from population history: a cladistic analysis of the geographical distribution of mitochondrial DNA haplotypes in the Tiger Salamander, Ambystoma tigrinum. Genetics 140: 767–82.
Tunstall-Pedoe H, Kuulasmaa K, Amouyel P, Arveiler D, Rajakangas AM, Pajak A. 1994. Myocardial infarction and coronary deaths in the World Health Organization MONICA Project. Registration procedures, event rates, and case-fatality rates in 38 populations from 21 countries in four continents. Circulation 90: 583–612.
UNESCO. 1995. Subcommittee on Bioethics and Populations Genetics of the UNESCO International Bioethics Committee, Nov. 15, 1995. Report.
United Nations General Assembly. 1949. The universal declaration of human rights. New York,. King Typographic Service Corp. 14 p.
United Nations. 1976 International covenant on civil and political rights, official records, first meeting of states parties, Sept. 20, 1976. Decisions.
United Nations Commission on Human Rights. 1995. Mataatua declaration on cultural and intellectual property rights of indigenous peoples. Cultural Survival 20 (Summer): 52–53.
United States Congress. 1993. Testimony of Francis Collins, NIH, and David Galas, DOE. Human Genome Diversity Project. Hearing before the Senate Committee on Governmental Affairs, Senate, 103 Congress, first session, April 26, 1993. Washington, D.C.: US Government Printing Office. p 6–15.
Vigilant L, Stoneking M, Harpending H, Hawkes K, Wilson AC. 1991. African populations and the evolution of human mitochondrial DNA. Science 253: 1503–5.
Wade P. 1993. Blackness and race mixture in Colombia. Baltimore, Md.: The John Hopkins University Press.
Wang D, Sapolsky R, Spencer J, Rioux J, Kruglyak L, Hubbell E, Ghandour G, Hawkins T, Hudson T, Lipshutz R, Lander E. 1996. Towards a third generation genetic map of the human genome based on bi-allelic polymorphisms. Amer J Hum Genet 59:A3 (abstract).
Weber JL, Wong C. 1993. Mutation of human short tandem repeats. Hum Mol Genet 8:1123–28.
Weir R. 1995. DNA banking and informed consent. IRB: A review of human subjects research 17: 1–8.
Weng Z, Sokal RR. 1995. Origins of Indo-Europeans and the spread of agriculture in Europe: comparison of lexico-statistical and genetic evidence. Hum Biol 67: 577–94.

Williams-Blangero S. 1989. Anthropometric variation and the genetic structure of the Jirels of Nepal. Hum Biol 61:1–2.

Wolpoff MH. 1989. Mitochondrial evolution; the fossil alternative to Eden. In: Mellars P, Stringer C, editors. The human revolution. Princeton, N.J.: Princeton University Press. p 62–108.

[WMA] World Medical Association. 1964. Declaration of Helsinki: recommendations guiding medical doctors in biomedical research involving human subject, adopted by the 18th World Medical Assembly. Helsinki, Finland. Reprinted in Medical Ethics Declarations. World Med J. 1984 31:4.

Wu D. 1990. Chinese minority policy and the meaning of minority culture: the example of Bai in Yunnan, China. Human Organization 49:1–13.

Zhao TM, Lee TD. 1989. Gm and Km allotypes in 74 Chinese populations: a hypothesis of origin of the Chinese nation. Hum Genet 83: 101–10.

Zhao TM. 1996. A Mongoloid-Caucasoid mixed population Uyghur: genetic evidence and estimates of Caucasian admixture in the Chinese living in northwest China. Paper presented at International Conference on the Bronze Age Peoples of Eastern Central Asia, University of Pennsylvania Museum, April 19-21, 1996.

APPENDIX
A
Committee on Human Genome Diversity: Biographical Information

WILLIAM J. SCHULL (Chair) is currently the Director of the Human Genetics Center at the School of Public Health of the University of Texas Health Science Center in Houston, and Ashbel Smith Professor of Academic Medicine. Prior to joining the faculty of the University of Texas in 1972, he was a member of the Department of Human Genetics at the University of Michigan. He has been a visiting professor at such institutions as the University of Chicago, the University of Chile, the University of Heidelberg, and the John Curtin School of Medical Research at the Australian National University in Canberra. He has also served on numerous editorial boards, and committees of the US Public Health Service, National Academy of Sciences-National Research Council, National Institutes of Health, and the Department of Energy and its affiliated national laboratories. A substantial portion of his scientific career has been spent in Japan studying the effects of exposure to ionizing radiation on the survivors of the atomic bombing of Hiroshima and Nagasaki. As a result of this involvement he was awarded the Order of the Sacred Treasure by Emperor Akihito in 1992. His interests in the effects on health of exposure to ionizing radiation and the role of genetic factors in the etiology of common chronic diseases has resulted in over 400 articles in professional journals, including 14 books that he has either edited or written.

GEORGE J. ANNAS is the Utley Professor and Chair, Health Law Department, Boston University Schools of Medicine, Law, and Public Health. He is also the founder of the Law, Medicine and Ethics Program at Boston University. He has authored or edited a dozen books on health law and bioethics, including *The Rights of Patients, Judging Medicine, and Standard of Care*. He currently writes a regular feature on health law in the *New England Journal of Medicine*, is co-chair of the Medical Practice and Medical Research Committee of the American Bar Association (Section on Science and Technology), is a member of the Special Committee on Genetic Information Policy of the Commonwealth of Massachusetts, and is coauthor of the proposed Genetic Privacy Act (written for ELSI). His primary interests are civil rights and civil liberties, especially of patients and other vulnerable populations, protection of research subjects, international human rights law, and health law.

NORMAN ARNHEIM currently holds the Kawamoto Chair in Biological Sciences at the University of Southern California. Prior to arriving at USC in 1985 he was Chairman of the Human Genetics Department and Senior Scientist at Cetus Corporation (1983-1985) and a Professor of Biochemistry at the State University of New York, Stony Brook (1968-1983). Dr. Arnheim received his PhD degree from Dr. Curt Stern at Berkeley, was a postdoctoral fellow in Allan Wilson's laboratory and spent a sabbatical year with Ed Southern in Edinburgh. His current research interests center on germline genetic instability and involve the analysis of DNA sequences in single human and mouse sperm cells using the polymerase chain reaction. Dr. Arnheim has over 100 scientific publications and has lectured widely in the United States and abroad.

JOHN BLANGERO is currently an Associate Scientist in the Department of Genetics at the Southwest Foundation for Biomedical Research in San Antonio, Texas. He holds a BA from Youngstown State University and a PhD (1987) from Case Western Reserve University. His research interests include statistical genetics, genetic epidemiology, and anthropological genetics. He has published 70 articles in professional journals primarily focusing on statistical and theoretical aspects of human genetics. He works on a number of ongoing projects including the identification of genes influencing predisposition to heart disease, diabetes, and infectious diseases. His research on the genetic basis of normal human variation has taken him to many parts of the world including projects in Tibet and Nepal.

ARAVINDA CHAKRAVARTI received his PhD in human genetics from the Graduate School of Biomedical Sciences at the University of Texas Health Science Center in Houston in 1979. He was a postdoctoral fellow at the University of Washington in Seattle during 1979-1980, subsequent to which he joined the faculty at the Department of Biostatistics, University of Pittsburgh, as an Assistant Professor. He spent 13 years at Pittsburgh, eventually moving to the Department of Human Genetics as a Professor of Human Genetics and Psychiatry. Dr. Chakravarti is a past Associate Editor of the *American Journal of Human Genetics*, currently one of the Editors-in-Chief of *Genetic Epidemiology and Genome Research*, on the Advisory Editorial Boards of the *European Journal of*

Human Genetics, Human Molecular Genetics and *Cytogenetics and Cell Genetics*, and a member of the NIH Mammalian Genetics Study Section. He joined Case Western Reserve University in January 1994.

VIRGINIA R. DOMINGUEZ is Co-Director of the International Forum for US Studies, Professor of Anthropology, and immediate past Director of the Center for International and Comparative Studies at the University of Iowa. Nationally she serves on the Board of Directors of the Society for Cultural Anthropology and on the editorial boards of *Public Culture,* the *American Ethnologist*, Public Worlds Books (University of Minnesota Press), and Transnational Cultural Studies (University of Illinois Press). She was born in Cuba, but spent much of her early life in and out of the US Since completing her PhD in 1979 at Yale University, she has taught at Duke University, the Hebrew University of Jerusalem, the University of California at Santa Cruz, and the University of Iowa. Her work has focused for many years on the historical and cross-cultural analysis of systems of social classification, how they develop, become discursively naturalized and institutionally entrenched. Among her publications are 4 books, including *White by Definition: Social Classification in Creole Louisiana* (1986) and *People as Subject, People as Object: Selfhood and Peoplehood in Contemporary Israel* (1989), and 4 (co)edited collections, including *Questioning Otherness* (1995), *(Multi)Culturalisms and the Baggage of Race* (1995), and the forthcoming *From Beijing to Port Moresby: The National(ist) Politics of Cultural Policies.*

GEORGIA M. DUNSTON is Professor and Interim Chair of the Department of Microbiology at Howard University College of Medicine, where she has been on the faculty since 1972. She holds a BS from Norfolk State University, MS from Tuskegee University, and PhD in Human Genetics from the University of Michigan. She is currently a fellow in the Visiting Investigator's Program at the National Center for Human Genome Research (NCHGR) in the Laboratory of Gene Transfer, where she is conducting research on mutational analyses of the BRCA1 gene in African-American breast cancer families and participating in development of a collaboration between Howard University and the NCHGR in a research project on the West African origins of non-insulin dependent diabetes in African-Americans. In 1985, her interests in the biomedical significance of genomic polymorphisms in African-Americans led to her establishment of and service as the first and current director of the Human Immunogenetics Laboratory at Howard University, an NIH funded core research resource. She has served on the National Advisory Council for the National Institute of Environmental Health Sciences; the Genetic Basis of Disease Review Committee for the National Institute of General Medical Sciences, and as organizer of international workshops and symposia on human leukocyte antigen (HLA) polymorphisms in African-Americans. She has published several articles in professional journals and has been invited to speak on HLA polymorphisms in African-Americans and its relevance to clinical transplantation and disease susceptibilities at universities and conferences in the United States and abroad. Her interests in the biomedical implications of human genome variability is the basis of current efforts to build research collaborations with investigators at African universities and facilitate greater international collaboration between African and African-American scientists in genomic research in African-American pedigrees. The latter has resulted in current membership on the North American Committee for the Human Genome Diversity Project.

WARD H. GOODENOUGH is university professor emeritus of anthropology at the University of Pennsylvania, whose faculty he joined in 1949. His empirical research has been on the cultures and languages of Micronesia and Melanesia in the Western Pacific. He has also been concerned with how to put together ethnological, archaeological, linguistic, and human biological evidence to determine how the prehistoric settlement of the Pacific Islands came about. His theoretical interests have centered on the nature of culture and on the processes by which cultures come into being, are maintained, change, and evolve. His methodological interests have been on data gathering and analysis such as to arrive at replicable and testable models of the cultures of specific peoples (e.g., how to determine what one needs to know in order to conduct a search of title in the traditional land tenure system of a people under study). He is past president of the Society for Applied Anthropology and the American Ethnological Society and former member of the board of directors of the American Association for the Advancement of Science. He has served as editor of *American Anthropologist*. His books and monographs are: *Property, Kin, and Community on Truk* (1951, 1978); *Cooperation in Change; An Anthropological Approach to Community Development* (1963); *Explorations in Cultural Anthropology* (editor, 1964); *Description and Comparison in Cultural Anthropology* (1970); *Culture, Language, and Society* (1971, 1981); *Trukese-English Dictionary* (with Hiroshi Sugita, 1980 and 1990); *Prehistoric Settlement of the Pacific Islands* (editor, in press).

RICHARD R. HUDSON is currently Professor of Biological Sciences at the University of California at Irvine where he has been since 1988. From 1983-1988 he was a Staff Fellow and then Research Mathematician at the National Institute of Environmental Health Sciences. He holds a BA from the University of California at San Diego and a PhD from the University of Pennsylvania. He has served on the editorial board of the journal *Genetics* and the journal *Theoretical Population Biology*. His main interests are the theory of population genetics and its application to understanding molecular variation within and between species.

ERIC T. JUENGST is Associate Professor of Biomedical Ethics at the Case Western Reserve University School of Medicine in Cleveland, Ohio. He received his PhD in Philosophy from Georgetown University in 1985, and has taught medical ethics and the philosophy of science on the faculties of the medical schools of Penn State University and the University of California, San Francisco. His research interests and publications have focused on the conceptual and ethical issues raised by new advances in human genetics and biotechnology, and from 1990 to 1994, he was the first Chief of the Ethical, Legal and Social Implications Branch of the National Center for Human Genome Research at the US National Institutes of health. He currently serves on the National Ethics Committee of the March of Dimes and the editorial boards of the *Journal of Medicine and Philosophy, Human Gene Therapy,* and the *American Journal of Medical Genetics.*

MICHAEL M. KABACK, MD, is a Professor in the Departments of Pediatrics and Reproductive Medicine, and Chief, Division of Medical Genetics, Department of Pediatrics, University of California, San Diego. He is the Director of the State of California Tay-Sachs Disease Prevention Program and of the International Tay-Sachs Disease Testing, Quality Control, and Data Collection Program. He received a BA from Haverford College and an MD from the University of Pennsylvania, After training and fellowships (Johns Hopkins and National Institutes of Health) he has held faculty positions at Hopkins, UCLA, and UCSD. From 1986-1992, he served as Chair, Department of Pediatrics, UCSD, and Pediatrician-in-Chief, Children's Hospital and Health Center, San Diego. Dr. Kaback is a Founding Board Member of the American College of Medical Genetics. He has served as Vice President of the Society for Pediatric Research, President of the Western Society for Pediatric Research, and President of the American Society of Human Genetics. Dr. Kaback is a member of the Institute of Medicine, National Academy of Sciences, and a Fellow of the American Association for the Advancement of Science. He is a recipient of the 1993 William Allan Award, from the American Society of Human Genetics, for outstanding contributions to human genetics. Dr. Kaback is the North American Editor for *Prenatal Diagnosis* and Associate Editor for *Advances in Pediatrics.* He is a national and international lecturer on various topics in medical genetics. He has authored or co-authored 8 books on these topics and has co-authored more than 200 publication in the biomedical literature. Dr. Kaback's major research interests include genotype-phenotype correlation in the lysosomal storage disorders, applications of genetic technologies to the control of human genetic disease, and psycho-social issues in population-based genetic screening.

DANIEL R. MASYS is Director of Biomedical Informatics, Office of the Dean, UCSD School of Medicine and Associate Professor of Medicine. An honors graduate of Princeton University and the Ohio State University College of Medicine, he completed postgraduate training in Internal Medicine, Hematology and Medical Oncology at the University of California, San Diego, and the Naval Regional Medical Center, San Diego. He served as Chief of the International

Cancer Research Data Bank of the National Cancer Institute, National Institutes of Health, and from 1986 through 1994 was Director of the Lister Hill National Center for Biomedical Communications, which is the Research and Development Division of the National Library of Medicine. At the Lister Hill Center, Dr. Masys was the principal architect of a program in computational biology which became the National Center for Biotechnology Information. He also served as the NIH representative to the federal High Performance Computing, Communications, and Information Technology Committee, which advised the President's Office of Science and Technology Policy in the area of advanced computing and National Information Infrastructure. Dr. Masys is a Diplomate of the American Board of Internal Medicine in Medicine, Hematology, and Medical Oncology. He is a Fellow of the American College of Physicians and a Fellow of the American College of Medical Informatics. He is a founding associate editor of the Journal of the American Medical Informatics Association, and has received numerous awards including the NIH Director's Award, Public Health Service Outstanding Service Medal, and the US Surgeon General's Exemplary Service Medal.

KATHRYN L. MOSELEY is the Director of Biomedical Ethics for the Henry Ford Health System. She is board certified in both pediatrics and neonatology. Her research interests include ethnic and cultural issues in medical ethics as they affect patients and providers. Dr. Moseley received her undergraduate degree from Harvard University and her medical degree from the University of Michigan. She studied moral theology at Aquinas Institute of Theology in St. Louis. During that time, Dr. Moseley served on the faculty of the Center for Health Care Ethics at the St. Louis University School of Medicine. She also completed a fellowship in clinical medical ethics at the Center for Clinical Medical Ethics at the University of Chicago. Dr. Moseley is a board member of the Society for Bioethics Consultation and is the President of the Medical Ethics Resource Network of Michigan. She is also a member of the Committee on Bioethics of the American Academy of Pediatrics.

ROBERT R. SOKAL, is currently Distinguished Professor Emeritus of Ecology and Evolution at the State University of New York, Stony Brook. His background is population biology, anthropology, and systematics. Education: St. John's University, Shanghai, China, BS (biology), 47; Univ. of Chicago, PhD (zoology), 52. His professional experience includes instructor to associate professor, University of Kansas, 51-61; professor, 61-69; professor, State University of NY at Stony Brook, 68-72; leading professor 72-91; distinguished professor 91-95, 95-. He held concurrent positions as Fulbright visiting professor, Hebrew and Tel Aviv Universities, Israel, 63-64; visiting distinguished scientist, University of Michigan, 75-76; visiting professor, Institute for Advanced Studies, Portugal, 71-80; visiting professor, University of Vienna, Austria, 77-78; Fulbright professor, University of Vienna, Austria, 84; visiting professor, College de France, Paris, 89; editor, American Naturalist, 69-74; president, Classification Soc, 69-71; president, Society for the Study of Evolution, 77; director, NATO Advanced Study Institute, 82; president, American Society of Naturalists, 84; president, International Federation of Classification Society, 88-89.

ALAN R. TEMPLETON is currently Professor of Biology and Genetics at Washington University, St. Louis, Missouri. He first went to Washington University in 1977 as an Associate Professor, and was promoted to Professor in 1981. Between 1974 to 1977, he was an Assistant Professor of Zoology at the University of Texas at Austin. He holds a BA from Washington University in Zoology, an MA in Statistics from the University of Michigan, and a PhD in Human Genetics from the University of Michigan. He has served as a visiting professor at the University of Sao Paulo (Brazil), the University of Michigan, and Oxford University (Merton College). He has served on the editorial board of six journals, and was an editor of *Theoretical Population Biology* from 1981 to 1990. He is a member of several scientific organizations, and is currently the President of the Society for the Study of Evolution. He was one of the founding members of the Society for Conservation Biology, and he is active in many organizations and on boards related to conservation issues. He has published 128 article since 1973 on his research in population genetics, human genetics, conservation biology, and speciation. Since 1973 he has given 126 invited seminars at universities in the United States and abroad, and has given numerous presentations at research conferences and 74 invited symposia talks.

LAP-CHEE TSUI is Senior Scientist and Sellers Chair of Cystic Fibrosis Research in the Department of Genetics at the Research Institute of the Hospital for Sick Children, Toronto, and Professor of Molecular and Medical Genetics at the University of Toronto. He was born in Shanghai, raised and educated in Hong Kong, and there awarded his bachelor and master degrees from the Chinese University of Hong Kong. In 1979, he received his doctoral degree (PhD) from the University of Pittsburgh. After training briefly in the Biology Division of Oak Ridge National Laboratory, he joined the Department of Genetics at The Hospital for Sick Children to work on cystic fibrosis (CF), a frequently fatal inherited disorder that affects about 1 in 2,500 Caucasian children in the world. In 1985, together with Dr. M. Buchwald and scientists at Collaborative Research Inc., he identified the first DNA marker linked to CF on chromosome 7. Four years later, Dr. Tsui, together with Dr. Jack Riordan at Toronto's Hospital for Sick Children and Dr. Francis Collins at University of Michigan, isolated the Cystic Fibrosis Transmembrane Conductance Regulator (CFTR) gene. His current research activities include molecular genetics of CF, a human genome project on the study of chromosome 7, and genetic studies in Tourette syndrome. Dr. Tsui has published 160 peer-reviewed articles and 60 invited papers. He has served on many national and international advisory committees and is currently on the editorial board of 8 scientific journals. He has received numerous awards for his research and 4 honorary degrees.

GEORGE C. WILLIAMS is Professor Emeritus of Ecology and Evolution at the State University of New York, Stony Brook, and Adjunct Professor of Biology at Queen's University, Kingston, Ontario. He has a PhD in zoology from UCLA and Sc.D. (honorary) from Queen's University. He has been Instructor and Assistant Professor at Michigan State University, Distinguished Visiting Investigator at the Museum of Zoology, University of Michigan, Fellow of the Center for Advanced Study in the Behavioral Sciences, a Guggenheim Fellow, and Visiting Investigator at the Marine Research Institute, Reykjavik. He was elected Ecologist of the Year (1989) by the Ecological Society of America.

TANIA WILLIAMS is a program officer in the Commission on Life Science's Board on Biology. Ms. Williams has worked on various projects for the Research Council since 1990. She received a Commission on Life Sciences staff performance award in 1992, a Commission on Geosciences, Environment, and Resources staff performance award in 1995, and a Commission on Life Sciences group performance award in 1995. She has staffed studies on environmental information for outer continental shelf oil and gas leasing decisions in Alaska, protection and management of salmon in the Pacific Northwest, wetland characterization, biomonitoring of environmental status and trends, scientific bases for preservation of the Mariana crow, and valuing biodiversity. Before she joined the Research Council, she was a cost-benefit analyst and technical writer for Wilson, Hill Associates and a research assistant in retail stock trends for Alex. Brown & Sons. She holds a BS in physiological psychology from Allegheny College.

APPENDIX B

Acknowledgments

The committee acknowledges with appreciation presentations made at meetings of the committee, personal communications, and respondents to its questionnaire.

Paul T. Baker, Pennsylvania State University (Professor Emeritus), Kaneohe, HI
Cynthia M. Beall, Case Western Reserve University, Cleveland, OH
Jaume Bertranpetit, University of Barcelona
Walter Bodmer, Imperial Cancer Research Fund, London, England
Anne Bowcock, University of Texas Southwest Medical Center, Houston, TX
Lisa Brooks, National Science Foundation,[1] Arlington, VA
Beth Burrows, Edmonds Institute, WA
Antonio Cao, Universita Degli Studi Di Cagliari, Cagliari, Italy
L. Luca Cavalli-Sforza, Stanford University School of Medicine
Susan Cheng, Roche Molecular Systems, Alameda, CA
Mary Clutter, National Science Foundation, Arlington, VA
Andrew G. Clark, Penn State University, University Park, PA
Christene deRaadt, New Zealand
Robert E. Ferrell, University of Pittsburgh, PA
Charles A. Gardner, National Institutes of Health, Bethesda, MD
Stanley M. Garn, Center For Human Growth and Development, Ann Arbor, MI
Clifford Geertz, Princeton University, NJ

---

[1] Dr. Brooks is now with the National Institutes of Health.

Takashi Gojobori, National Institute of Genetics, Mishima, Japan
Betty Graham, National Insititutes of Health, Bethesda, MD
Henry T. Greely, School of Law, Stanford University, Palo Alto, CA
Roger C. Green, University of Auckland, New Zealand
Judith Greenberg, National Institutes of Health, Bethesda, MD
Henry Harpending, Penn State University, University Park, PA
Russell Higuchi, Roche Molecular Systems, Alameda, CA
W.W. Howells, Kittery Point, ME
The Honorable Daniel K. Inouye, US Senate
Ulf Landegren, Uppsala University, Sweden
Kenneth K. Kidd, Yale University, New Haven, CT
Mary-Claire King, University of Washington, Seattle
Ruth Liloqula, Ministry of Agriculture and Fisheries, Solomon Islands
Darryl Macer, University of Tsukuba, Tsukuba Science City, Ibaraki, Japan
Richard S. MacNeish, Andover Foundation for Archaeological, Research, Andover, MA
Partha P. Majumder, Indian Statistical Institute, Calcutta, India
Victor McKusick, Johns Hopkins University, Baltimore, MD
Aroha Te Pareake Mead, Aotearoa, New Zealand
D. Andrew Merriwether, University of Pittsburgh, PA
John Moore, University of Florida, Gainesville
John Moore, Edmonds Institute, WA
Newton E. Morton, University of Southampton, UK
Yolanda T. Moses, American Anthropological Association, Arlington, VA
Arno G. Motulsky, University of Washington, Seattle
Mary Margaret Overbey, American Anthropological Association, Arlington, VA
Sergio D.J. Pena, Universidade Federal de Minas Gerais, Belo Horizonte, Brasil
Henry Pitot, University of Wisconsin, Madison
Helen Ranney, San Diego, CA
Dai Rees, European Science Foundaton, London, England
Jeffrey Rogers, Southwest Foundation for Biomedical Research, San Antonio, TX
Charles Scriver, Montreal Children's Hospital and McGill University, Canada
Ron Sederoff, North Carolina State University, Raleigh
Hope J. Shand, Rural Advancement Foundation International, Pittsboro, NC
Anthony D. Socci, Coordination Office of the US Global Change Research Program, Washington, DC
Abadio Green Stocel, Organización National Indígena de Colombia, Bogotá
Leonor Zalabata Torres, Community Leader and Representative of the Arhuaco people, Sierra Nevada de Santa Maria, Colombia
Evon Z. Vogt Jr., Peabody Museum, Cambridge, MA

Susan Wallace, Human Genome Organization, Bethesda, MD
Kenneth M. Weiss, Penn State University, University Park, PA
Mark Weiss, National Science Foundation, Arlington, VA